Jürgen Kletti (Ed.)

Manufacturing Execution Systems - MES

Jürgen Kletti (Ed.)

Manufacturing Execution Systems – MES

With 100 Figures

Editor

Dr.-Ing. Jürgen Kletti (Ed.)

MPDV Mikrolab GmbH
Römerring 1
74821 Mosbach
Germany

j.kletti@mpdv.de

ISBN 978-3-642-08064-7 Springer Berlin Heidelberg New York
e-ISBN 978-3-540-49744-8

Springer is a part of Springer Science+Business Media

springer.com

Cover: Frido Steinen-Broo, eStudio Calamar, Spain

Printed on acid-free paper 68/3100/YL - 5 4 3 2 1 0

Foreword

The transformation of the classic factory from a production facility into a modern service center has resulted in management problems for which many companies are not yet prepared. The economic efficiency of modern value creation is not a property of the products but rather of the process. What this means is that the decisive potentials of companies are to be found not so much in their production capability but in their process capability.

For manufacturers the requirement for process capability, which has in the meantime become the basis of the certification codes, gives rise in turn to the requirement that all value-adding processes be geared to the process result and thus to the customer. A necessary condition of process transparency is the ability to map the company's value stream in real time, without the acquisition process involving major outlay – a capability which is beyond the dominant ERP systems.

Today modern manufacturing execution systems (MES) can offer real-time applications. They generate current and even historical maps for production equipment and can thus be used as a basis for optimization processes. As early as the beginning of the 1980s work started on methods of this kind which were then known as production data acquisition or machine data collection. But while the main emphasis in the past was on achieving improvements in machine utilization, today the concern is predominantly to obtain real-time mapping of the value stream (supply chain).

Here the increasing complexity of production calls for a holistic view of production and services equipment and facilities: detailed planning, status detection, quality, performance analysis, material tracking and so on have to be registered and displayed in an integrated manner.

In the mid-1990s the concept of the manufacturing execution system developed in the USA from out of these exigencies. A non-profit organization with the name of MESA (Manufacturing Execution System Association) started standardizing these applications and thereby raised three application layers of a production facility into a principle. MESA defines the level of production itself, the level of production management (in other words, MES), and the level of corporate management.

Further works of standardization relating to this subject area are already in the process of development. Accordingly, an ISA S95 has been approved, whereby NAMUR, an association of process manufacturers, has come up with its own guideline for its own particular area of manufacturing.

Very recently too the Verein Deutscher Ingenieure (VDI) has picked up this topic and is working on issuing a guideline tailored to the particular concerns of European manufacturers.

The expectations placed upon a manufacturing execution system for increasing performance are correspondingly high. The practitioner will have a particular interest in topics such as TQM, SIX Sigma, operations planning or optimized material movements.

Even today the growing interest in this area is indicated by the increasing use of the term "MES" in specialist literature and market surveys not to mention in work on standardization in which a number of committees are involved.

The term MES should be methodically systematized in order to give manufacturers the broadest overview possible of the capabilities and the different practical possibilities of an MES and thus put them in a position of exploiting this overview to orient themselves within the broad supply of goods to the market. In the present book, experienced specialists in the field throw detailed light on different aspects of an MES without which it is not possible to run a modern company profitably today.

Gaining control of the processes is more and more becoming a central requirement for companies if they are to be able to produce profitably even in a location such as Germany.

Professor Johann Löhn
President of the Steinbeis-Hochschule Berlin
Government commissioner for technology transfer
Baden-Württemberg

Table of contents

1 New ways for the effective factory

1.1 Requirements for tomorrow's manufacturing

The classic factory has been defined by its manufacturing of goods. The goods and their value have been measured primarily by their material components. This is no longer adequate today. Increasing globalization is necessarily leading towards more anonymous products out of long supply chains and with an increasing complexity to track their origins. This implies a shifted focus from control of production creation (vertical integration) to control of product perception by the customer (OEM). Customers today take it for granted that products will be of first-class quality. Anyone wishing to stand out from the competition in the future needs a strategy which offers the customer an additional added value, such as, for example, high flexibility, short delivery times, high delivery reliability, wide range of variants, shorter product life cycles – properties which are not created by production but by the processes. The term "adaptive manufacturing", which is heard more and more often these days, describes this approach as "connecting the machines to the markets."

For this reason many classic manufacturers today already define their production facilities as a service center, thereby signaling to the customer that they understand the processing of material into a finished product as also being a service for the customer. This increase in closeness to the customer initially results in cost increases. Modern producers attempt to cancel out these increased costs by rethinking their vertical integration, in some cases by using standard components or by sourcing suitable components on the global market. The modern producer is thus faced with forces which can be referred to as networking, dynamization and individualization .

The term "networking" means the increasing inter-company cooperation. In today's public discussion is principle is named globalization. Thanks to this networking the manufacturer can purchase on the market the components he needs thus leaving him able to concentrate on his core competences which he then, in a supply chain management strategy, incorporates effectively into the total product manufacturing chain.

Dynamization originates in strong market fluctuations which, driven by more information and ever more rapidly disseminated information, encourage customers to make rapid changes in their purchasing habits. The ever faster turning wheel of technological development makes a further contribution to these effects. Errors are more likely in complex, collaborative processes than in simple, closely coupled processes. The failure management resulting from this and also the frequent and faster changes in customer orders further stimulate the dynamics.

The change towards buyer's markets and greater focusing on the customer demands more individualization from manufacturers. What the customer wants is a product which is tailored to his requirements. The logical consequence is an increase in the range of variants which the producing company must offer to its customer.

Networking, dynamization and individualization create increased risks and complexities in the production facilities and demand that producers be capable of change. This turbulence is characterized by new requirements in the internal processing of orders and also in external market dynamics. These changed requirements are characterized by stronger external networking, by collaboration with multiple and/or with new partners, and by faster internal structural and technical adaptations. This new process instability makes manufacturing close to an economic optimum more difficult and also fosters not only inefficient information management but also outdated business processes. The consequence for the customer is poor delivery reliability and lead times as well as unsatisfactory product quality. In many cases the manufacturer experiences long delivery times which in turn result in excessive inventory levels. The consequence is that more capital is tied up.

The list of effects generated by turbulence and changeability can be continued. These effects impact on every level in a manufacturing company, often in different ways with different effects. The consequences of these effects can be resisted by creating more transparency within the levels and between the levels, by improving reaction capabilities, and by securing cost-efficiency.

To increase transparency, there must be greater integration of the business processes affected. Impediments and obstructions which still exist today in communications between the corporate levels of management, production management and also the production department itself must be dismantled and removed. Information will need to flow faster and more effectively within the levels. The vertical integration or continuity from management to production so often required today should be supplemented by horizontal integration. Improved reaction capability will develop on this basis of increased transparency. Faster information means that problems and unplanned events will be detected faster. This makes faster reaction possible and remedial action can be taken faster. With these resources, production planning can be set up which is characterized by short reaction times and which thus earns its description as a fine planning control system with short control cycles.

With this set of tools, deliveries or services can be modified at short notice in an economic and cost-oriented manner thereby complying with customer for flexibility. But the efficient introduction of changes, a satisfactory level of adaptability to the changing needs of the company, and the ability of existing technologies and systems to be readily integrated also have to be developed and refined in a manufacturing company.

The potential benefits which emerge from these elements – such as better customer service due to improved delivery reliability and delivery capability, as also product quality and information capabilities, cost savings due to inventory reductions, improved workforce position, motivation arising from control of production, and so on – supply decisively important key performance indicators for the current competitive environment.

These three elements in an improvement process – transparency, responsiveness and cost efficiency – have been partially put into practice in industry in recent years. Some advances have been made here, particularly on the level of corporate management. In the commercial departments of corporate management, changes do not come into effect in seconds, minutes or hours but rather in days, weeks or months. The situation is completely different in the field of production management and automation. Many short-term activities are required here and they in turn call for tools that support online ad hoc decisions. Every minute of downtime for a machine or part of a plant costs money. Every minute of these production problems eats into profits. In cases of this kind it is very easy to demonstrate a clear relationship between the benefits and the costs of measures and tools for preventing or reducing breakdowns.

Today, particularly in production management, the aim of "increasing transparency, responsiveness and cost efficiency" means that new paths must be taken and increased effort applied to measures which have already been introduced. One tool which supports these objectives is the so-called MES system (manufacturing execution system). MES is a method that has developed from what tend to be the classic disciplines, such as production data acquisition, staff work time logging, quality assurance and finite scheduling. The homogenized and compacted version of these techniques can be grouped under the heading MES. The aim of an MES is to make the value-adding processes transparent and on the basis of this transparency to create not only horizontal but also vertical control cycles. The cycle time of these control cycles will depend on the tasks being performed and, to take the example of production, be measured not in days or shifts as is normal in a traditional ERP environment but rather in multiples of minutes. In this way production can react quickly and cost-effectively to meet new requirements.

The present book is intended to throw light on various aspects of MES and the use of MES and should also describe how potentials for improvement can be identified and exploited, even in a heavily automated industry.

1.2 Production structures

The aim of achieving an improvement in economic efficiency is not, of course, a new demand but is rather a permanent process which has increasingly challenged manufacturing industry over the last few decades. In the media we only hear about particularly major pushes made in this direction (such as the case of Ignazio Lopez or jobs being exported from Germany). Alongside the improvement in processing technology and the reduction in material and labor costs, this striving for greater efficiency has initially been met by an improvement in production structures and control methods with the aim of improving the passage of an order through production. For this reason new approaches have been developed in recent years which satisfy demands for shorter lead times and greater flexibility, particularly with regard to the increasing number of product variants. Some of these production structures and control methods will be dealt with briefly below.

1.2.1 Orientation towards metrics

Different factors must be taken into account in the selection of suitable production structures. One important criterion is the production quantity planned. Due to the high degree of automation, the line structure delivers the highest productivity but recouping the high investment costs necessitates producing large quantities over long periods. Other important criteria are flexibility with respect to product change, production of variants, volume changes, work in progress inventory, working conditions, and so on. Here it is necessary to identify the maximum benefits by making an evaluation of the various structures with regard to these criteria.

Shop production

In shop production all machines which carry out the same tasks are grouped together into shops – for example, all lathes and turning machines in the turning shop, all milling machines in the milling shop (layout by machine). In this case the time flow of production is tied to batches. Not until the last workpiece in a batch has been processed are all members of the batch sent on to the next operation. The result of this in a multi-step process is an unclear flow of material with long transportation paths, queuing and waiting times, large work in progress inventories and poor compliance with scheduled times. Shop production originated in the pursuit of high flexibility and simplified layout planning.

Production in decentralized structures

Product- or customer-oriented organizational units are grouped into decentralized structures which include several production stages (the factory in the factory). The objective is to combine the cost and productivity benefits of line or continuous production with the high flexibility of shop production. The assumption underlying the decentralized structures approach is that it is easier to coordinate small units since all of the units needed to carry out the task are grouped together in one area. Decentralized structures can thus be intensively aligned to specific competitive strategies.

Line flow production

Here machines and operations are organized in accordance with the order in which a product is processed (layout by product). Due to the fine time coordination and interlinking required for the individual operations (cycle timing) this structure is very sensitive to breakdowns as well as inflexible product variations. In addition, installations of this kind have high investment costs which is why they can be used economically only in large-scale production. Here line flow production does, however, offer the greatest productivity advantages over other production structures since queuing and waiting times, work in progress inventories as well as transportation paths are minimized.

1.2.2 Control methods

The selection of suitable control methods depends to a considerable extent on the production structure (for example, shop production or flow production). However, the type of orders to be processed (for example, customer or stock order, quantities, number of variants, spread of the orders, and so on) also plays an important role. In principle a distinction can be drawn here between the push principle and the pull principle.

Push principle

The push principle means that production orders are generated in a central production planning and control facility and are then executed in the production department. Examples of push methods of this kind are:

- MRP II (manufacturing requirement planning)
  The MRP II method developed out of MRP I (material requirements planning) by the incorporation of manpower and machine capacities in calculations. It is used primarily in normal and small series production on the shop principle since multistage production structures require an increased level of planning.
- Cumulative number concept
  With the cumulative number concept, material movements are recorded cumulatively over time (actual CN) with the aid of a cumulative number (CN) and compared with the planned value (target CN). Use of the cumulative number concept requires large production quantities and a linear production structure. For this reason this method is primarily suitable in full-scale and mass production with line or flow production.
- Load-dependent order release
  Load-dependent order release was developed in particular for one-off and quantity production on the shop principle of products which have a large number of variants. It regards machines as hoppers whose fill levels (number of orders) are controlled.

Pull principle

With the pull principle, items are only produced in response to a customer demand for them. The customer order generates a requirement in the final assembly department. This requirement in turn generates a requirement in pre-assembly, and so on – in other words, the sales order works its way backwards through production until it reaches material procurement. The aim of the pull principle is to reduce the control overhead and to make production more transparent and less inventory-hungry. Examples of pull methods of this kind are:

- Kanban
  The kanban method is based on autonomous control cycles between a consuming station and a producing station. Here the producing station receives a signal which tells it what parts are needed in what quantity at what time by the consuming station. The signal is given by means of kanban cards. Kanban is used predominantly in mass production on a flow production basis.
- CONWIP (constant work-in-process) is based on the kanban system but still includes the control cycles of several stations found in flow production.
- Synchronous production
  In synchronous production the ideal production line produces with the same work cycles as the customer or in accordance with customer call-

offs. Chaining the work steps means that it is only necessary to control a single process step in the entire process chain. The pacemaker process is the process which is directly controlled by the customer. The aim of this method is to achieve a continuous flow (one-piece flow).

– Agent control
  Higher level IT systems determine star dates on the basis of the customer dates. On the basis of this information, work pieces, installations and transportation systems negotiate the process flow decentrally and autonomously while taking the current state of production into account at all times.

1.2.3 Combinations of production structure and control method

As has already been said, not every control method is suitable for every production structure. In practice the following combinations are encountered:

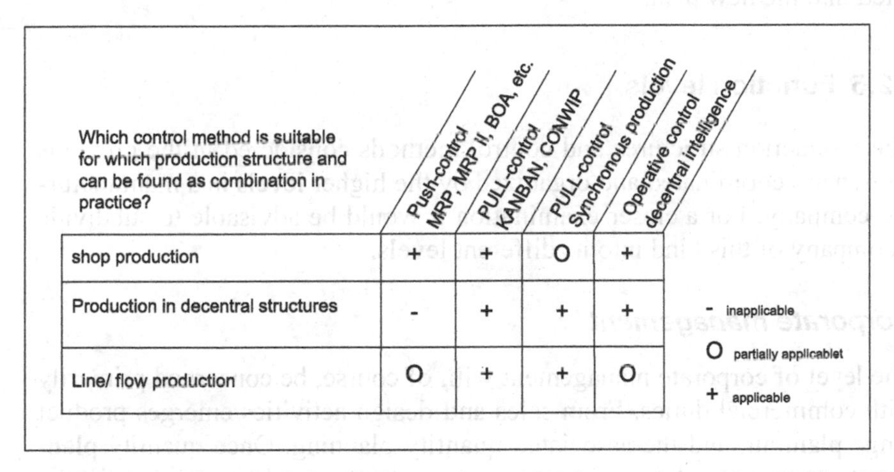

Which control method is suitable for which production structure and can be found as combination in practice?	Push-control MRP, MRP II, BOA, etc.	PULL-control KANBAN, CONWIP	PULL-control Synchronous production	Operative control decentral intelligence
shop production	+	+	O	+
Production in decentral structures	-	+	+	+
Line/ flow production	O	+	+	O

- inapplicable
O partially applicablet
+ applicable

Fig. 1.1. Control methods in relation to the production structure (Fraunhofer IITB, 2005)

1.2.4 Weaknesses in traditional PPS systems

Despite some sophisticated control methods, traditional production planning and control has serious weak points in the planning and scheduling of production orders. This is why we can see a trend towards pull approaches. These weak points include:

– Planning with uncertain planning input data (processing times, machine utilization, etc.)
– Planning split too coarse due to planning by the week or at best by the day
– Planning without an updated load horizon
– Missing or excessively late feedback about progress towards completion, faults and so on; this means only delayed control is possible
– Inflexible as regards rush orders or changes in requirements or dates
– Does not take actual capacity utilization into consideration

In the PPS strategies a tendency towards planning can be seen – in other words, towards a one way street principle without feedback. Transparency and responsiveness are thus not achieved. In principle an improvement process involving better planning must end at a specific point. Without real-time confirmation of completion, the control cycle consisting of the production plan and production itself will in the best case be run through once a day since inputs first need to be checked, corrected and incorporated into the new plan.

1.2.5 Function levels

The production structures and control methods considered in the previous section are coordinated and organized by the higher levels in a manufacturing company. For a closer examination it would be advisable to subdivide a company of this kind into its different levels.

Corporate management

The level of corporate management will, of course, be concerned primarily with commercial duties. From sales and design activities emerges product range planning and the associated quantity planning. Once quantity planning has been completed on a customer-, order- or stock-oriented basis, the order release will be given. As a result of this, or even dependent on it, the time scheduling and capacity requirements planning must start for production. In virtually all cases this planning step will be rough planning – in other words, using a rough grid commensurate with the processing time period, the capacities available are examined and also the units to be manufactured on these capacities. On the basis of the information flowing back from production the inputs for the next production period or for the next planning section can be changed if necessary.

Production management

Production management receives the order loading and the corresponding dates from corporate management and carries out sequencing and loading planning. This planning step should be referred to as "finite planning". Here the orders or operations are scheduled out to the available capacities with the most exact start dates possible being determined and passed on to the actual production department. This production management level also includes collection of the production data with whose help a real-time target/actual comparison can be carried out between input requirements and the real information.

All types of resource management are normally carried out on this level. The preparation of personnel deployment plans is a special task which is usually performed by production management. Even quality assurance with its wide range of functions regarding data acquisition and evaluation is normally a task which falls under production management.

The production level (automation level)

Machine and system control and also stock control are now assigned to the actual production department. In the same way, transportation control, maintenance and the actual manufacture of goods are tasks for the production level. Further on in this book this level will also be frequently referred to as the automation level.

Within the context of the present book, examination of production management plays a central role. This is where flows of material and of information intersect decisively in a manufacturing company. Production management also makes a substantial contribution to value creation. At this point unsuitable mechanisms may mean that not only no money is earned but also burned.

Production management determines the logistical performance of a company, particularly regarding its responsiveness to market influences. More recent control methods tend to be decentralized and responsibility delegated to lower hierarchy. In this way production management gains more and more responsibility and importance. Inter-company networking in the supply chain management environment takes place today more and more often on the level of actual production or of production management. Within this book this particular arrangement of levels is intended to serve as a model for all types of production departments.

1.2.6 Types of production

Three different so-called production types are to be distinguished. These are: discrete or shop production, process-line production or mass production, and the make-to-order manufacturer or plant and equipment manufacturer. The distinction is important at this point as further on in the book. We intend to show how these types of production require MES functionality. A brief summary of the basic characteristics of the three types follows.

Discrete or shop production

Here we have production orders from a series of operations which in some cases can be regrouped into assemblies. The discrete manufacturer would prefer transitions between his processing steps to be as short and smooth as possible. The availability of intermediate products is an important variable, as is organizing these intermediate products in interim storage facilities. Specific variables here are resource availability and above all flexibility in the processing of orders.

Process lines or mass production

The mass, process or line manufacturer links together his systems and machines to form lines which normally produce large quantities of a product. Flexible changes in order processing are only possible with qualifications. The fact that a line runs permanently is of central importance. Due to the complexity of installations, shifting orders to different resources is often impossible or if so, only to a limited extent. This means that a special logic will also have to be taken into account in production planning.

Make-to-order production/plant and equipment manufacturing

The make-to-order manufacturer or plant and equipment manufacturing typically has comprehensive bills of material which are often processed in manufacturing cells or in dedicated shops. These manufacturing cells have a certain amount of independence which means that sometimes there can be transitions between them which are not time-critical. Depending on the products being manufactured this kind of manufacturer may also have full-scale or small-series production facilities.

The relative closeness of deadlines shown in the diagram is intended to provide a qualitative representation of the different time horizons within which the three levels carry out their tasks. The range extends from long-term in ERP production range planning to virtually real-time or online on the automation level.

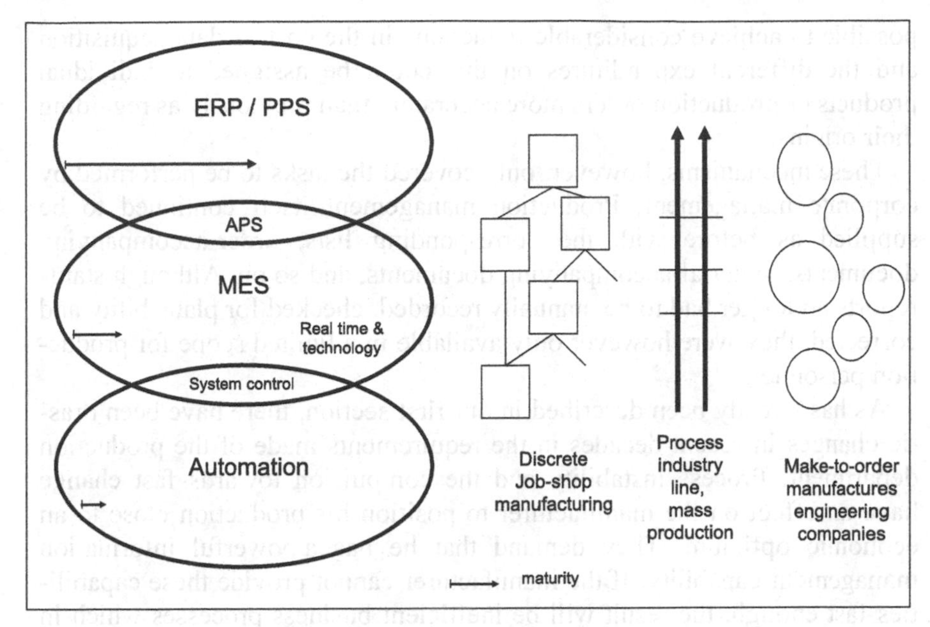

Fig. 1.2. The three principle types of production in manufacturing industry. Each of these types has an ERP, an MES and an automation level.

1.3 Classic IT support in production

In the early years of information technology, manufacturing enterprises were mainly "controlled" by commercially oriented systems. It was a gigantic step forward to automate classic manually oriented commercial services and to manage bookkeeping, inventory and orders received electronically. In the next stage of this process of automation it was possible to provide some of the above-mentioned production structures and control methods with support from so-called EDP systems. The milestones here were detailed planning of orders, resolution of orders into individual operations or sequences of operations, and the breaking down of products into individual assemblies. Supervisors in production were supplied with lists which contained sales planning information and the customer orders to be produced. Consumption of time, materials and other resources were reported back to the EDP system from production and this information recorded there – a complex method which was also likely to be encumbered with errors. Simpler and better was to send a status report once individual departments started to be equipped with dedicated data collection systems. In this way the EDP side was provided with a production data acquisition facility, the personnel department with staff work-time logging and the quality assurance department with a so-called CAQ system. It was possible to achieve considerable reductions in the cost of data acquisition and the different expenditures on this could be assigned to individual products or production orders more accurately than previously as regarding their origins.

These mechanisms, however, only covered the tasks to be performed by corporate management. Production management itself continued to be supplied as before with the corresponding lists, order-accompanying documents, material-accompanying documents, and so on. Although status reports no longer had to be manually recorded, checked for plausibility and corrected, they were however only available in a limited scope for production personnel.

As has already been described in our first section, there have been drastic changes in recent decades in the requirements made of the production department. Process instability and the compulsion towards fast change have an effect on the manufacturer to position his production close to an economic optimum. They demand that he has a powerful information management capability. If the manufacturer cannot provide these capabilities fast enough, the result will be inefficient business processes which in turn lead to poor on-time delivery performance, poor delivery times, unsatisfactory product quality, long lead times and excessively high inventory levels. In this regard, ERP/PPS systems have retained a high proportion of their old characteristics even until today. They do not, for example, support hierarchization and separation into levels as would be required in a production facility. The focus of all instances of optimization relates to planning – in other words, the "one-way street principle" and neglects control capability. Even now, the control cycles of an ERP system are longer than a shift while the production scheduler on the spot actually needs control cycles of the order of several minutes. Control cycles this fast are not to be found in an ERP-supported manufacturing organization. For this reason open control chains are predominant. The information flowing back from production is in some cases not available in a processed form until the next shift which means that it cannot be used as online information by the persons in charge in production.

Certain aspects of this problem area are defused into APS functionalities (Advanced Planning and Scheduling). Here the control cycles do not even last a week but can be reduced to one or two days. However even with APS we still are faced with the problem that control mechanisms within a shift or a day are only possible to a very limited extent and that the focus still remains on load planning and less on the control of production.

What we can compare with the APS functions are industry specific control stations. Here planning can be carried out not only almost in real time but also oriented to a certain extent to the technology. A typical control station takes into account aspects relating to the particular branch of industry such as, for example, the color sequence in injection molding or the suitability of tools and machine combinations for producing specific articles. But this control station approach still can only deliver a limited amount of improvement. If the current actual situation is not included in the corresponding new planning, we do not have a control facility here but rather a planning facility, as previously. If ERP-based planning is referred to as rough planning, the use of APS or control-station–oriented planning means that so-called detailed planning can be achieved.

More detailed and dedicated functionalities, such as online display of current states, display of utilization ratios, online interpretation of registered and unsatisfactory qualities, and also the display of incorrect states are missing in control station. Evaluations which tell you tomorrow what you could have done better today are only of interest in a statistical respectively historical view.

At this point even the term “transparency” takes on a new meaning. Transparency in modern production no longer only means the ability to comprehend past history without omission or gaps and from this to derive recommendations for future actions. Today, transparency also means visualizing realities in real time, drawing conclusions from this, and then communicating recommendations to the appropriate persons in order to put an immediate end to incorrect states.

1.4 Manufacturing Execution Systems (MES)

1.4.1 Emergence of the MES concept

The origins of the MES concept are to be found in the data collection systems of the early 1980s. The various disciplines in corporate management such as production planning, personnel, and quality assurance were furnished with dedicated data collection systems. This situation is shown in the following diagram: task areas which are almost mutually independent are equipped with special data collection systems.

With the rise of the CIM concept (Computer Integrated Manufacturing) a start was made on reproducing the interdependencies of these task areas in the IT systems as well. Production, personnel and quality were no longer seen as completely independent but rather data crossovers were permitted from one task to another. Unfortunately this approach, correct as it was in principle, did not emerge as a real and strong IT discipline. Trivialization of the problem definition and a misuse of the term by smaller system vendors in the sense that with time every data collection terminal

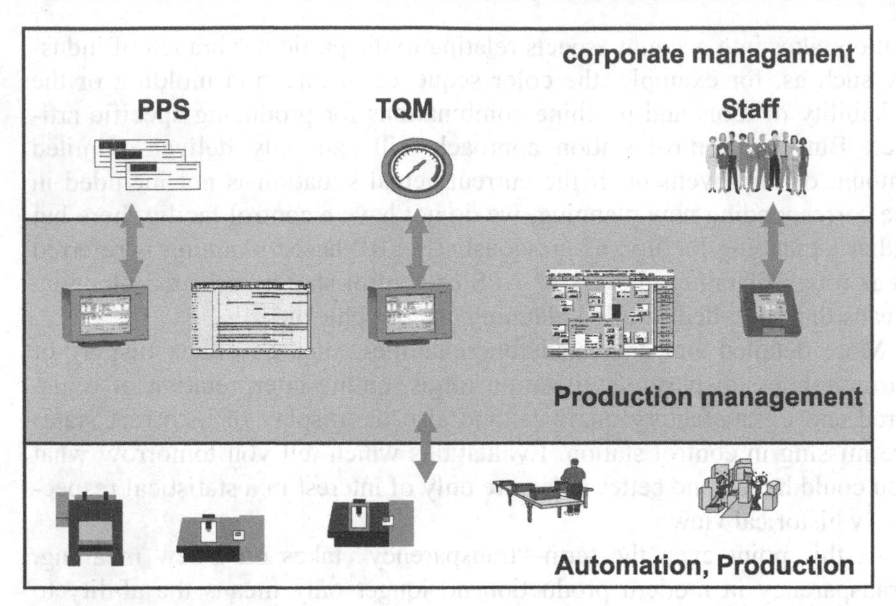

Fig. 1.3. Each area of activity in corporate management has a particular data collection method assigned to it which is also independent of the others

was labeled a CIM system. In this way CIM had spoiled its standardization potential as a problem-solving IT discipline for production.

In the early and mid-1990s the manufacturers of data collection systems commenced upgrading their in some cases specialized systems (labor time, PDA, CAQ, DNC, and so on) by adding features from associated fields (for example: staff work time logging with PDA, PDA together with MDE). With a small number of combination systems of this kind it was already possible to put together a data collection (and sometimes a data evaluation) system for many functional areas of a manufacturing company.

The system components were, however, independent of each other and synchronizing them required major work on interfacing. Over the course of time three groups of data collection/evaluation systems formed. From the independent data collection systems, combination systems emerged, some of which performed several tasks. All in all, the functionality of these combination systems describes the functional scope of MES today:

- For production matters: PDA, MDE, DNC, control station;
- For personnel matters: staff work time logging, access control, short-term manpower planning;
- For quality assurance matters: CAQ, measured data acquisition.

In the real world of production these three task areas cannot be separated from each other. Production accordingly needs suitable personnel to

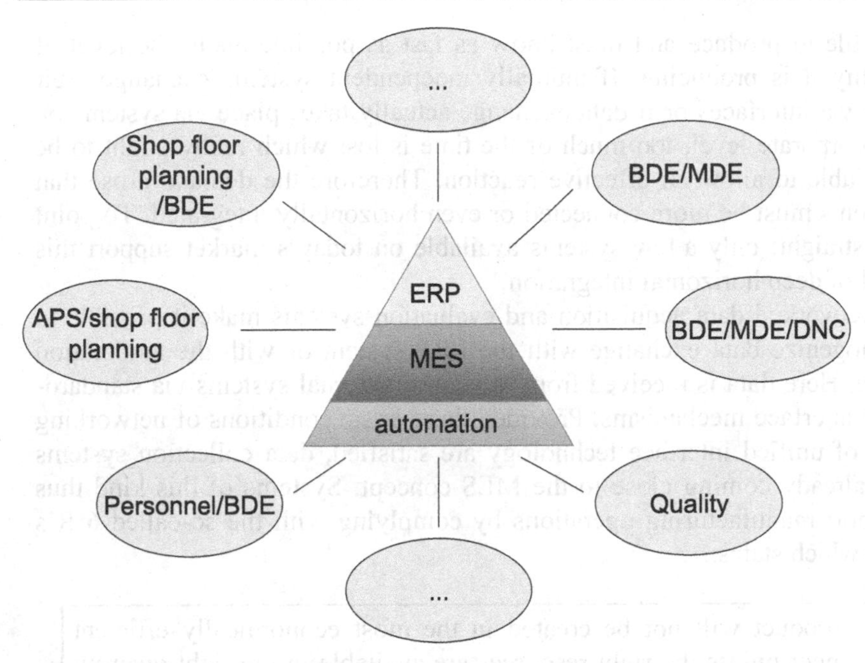

Fig. 1.4. MES integrates originally separate data collection systems

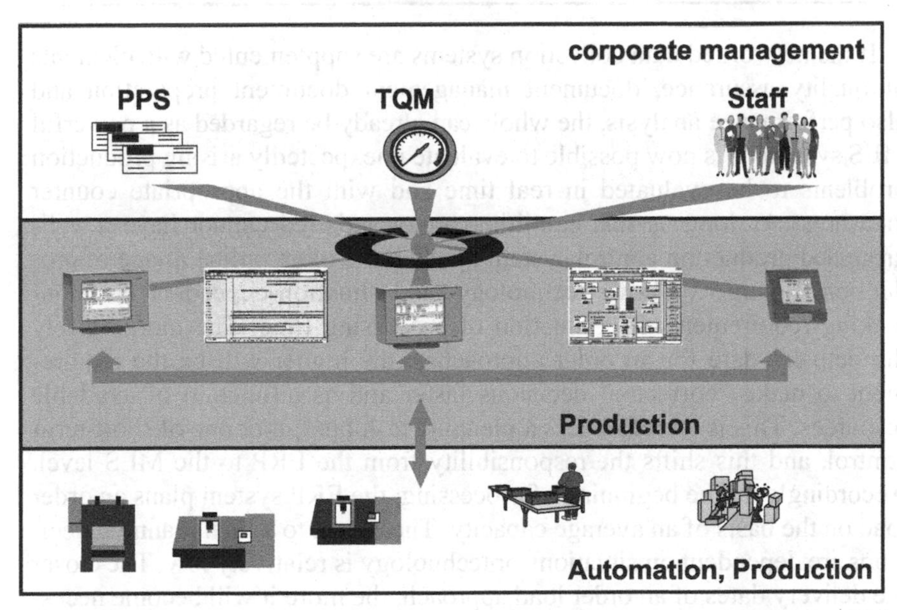

Fig. 1.5. Mutually independent data collection systems were networked and in some cases connected to corporate management and automation via uniform interfaces

be able to produce and must know as fast as possible about the level of quality it is producing. If mutually independent systems exchange their data via interfaces or if data exchange actually takes place via systems on the corporate level, too much of the time is lost which really ought to be available to allow an effective reaction. Therefore the demand arose that systems must be more connected or even horizontally integrated. To point out straight: only a few systems available on today's market support this kind of deep horizontal integration.

Networked data acquisition and evaluation systems make it possible to homogenize data exchange with the ERP system or with the automation level. Here data is received from or sent to external systems via standardized interface mechanisms. Provided these basic conditions of networking and of unified interface technology are satisfied, data collection systems are already coming close to the MES concept. Systems of this kind thus support manufacturing operations by complying with the so-called 6 R's rule which states:

> A product will not be created in the most economically efficient manner unless the right resources are available in the right quantity at the right place at the right time with the right quality and with the right costs throughout the entire business process.

If the networked data collection systems are supplemented with elements of quality assurance, document management document preparation and also performance analysis, the whole can already be regarded as a powerful MES system. It is now possible to evaluate unexpectedly arising production problems to be evaluated in real time and with the appropriate counter measures. As long as this condition is not satisfied cannot have a well-grounded production control system on the basis of an online image of production. Figure 1.6 shows technology- and situation-dependent decision-making requirements as a function of processing time. The more closely the delivery date for an order approaches, the higher will be the requirement to make "corrector" decisions faster and as a function of available resources. This is therefore less a planning task but rather one of short-term control, and this shifts the responsibility from the ERP to the MES level. Accordingly, at the beginning of processing, the ERP system plans an order load on the basis of an average capacity. The extent to which planning decisions are dependent on situations or technology is relatively low. The closer the delivery dates of an order load approach, the more it will become necessary to make adaptations in response to unforeseen problems. The closer the delivery date approaches the more this control function will be dependent on technology, situation and remaining capacities.

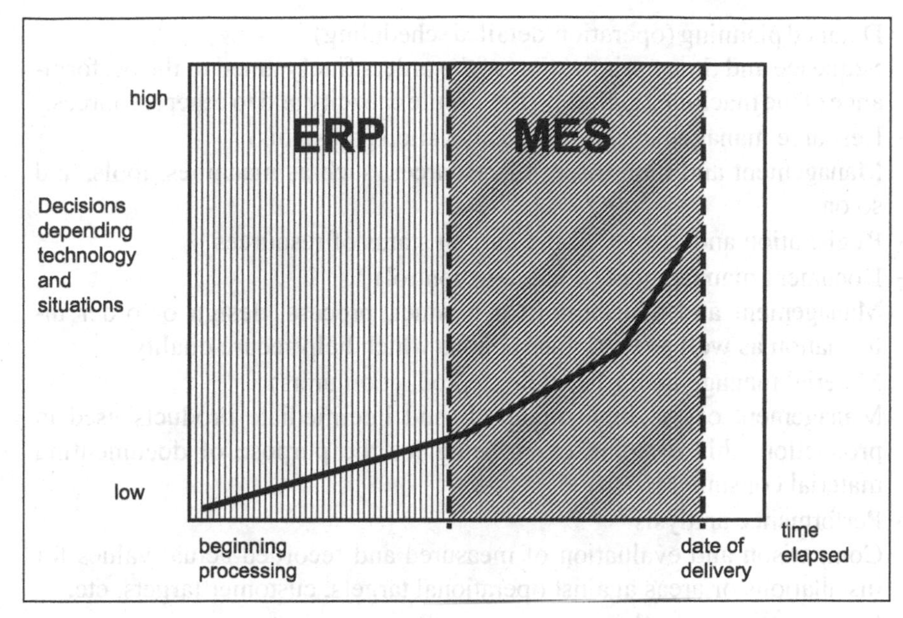

Fig. 1.6. Dependence of control requirements on planning time

1.4.2 Current standards

The subject of MES has been taken up by a number of institutions which are attempting, to protect the term MES against trivialization. Various implementations types may be found and in the present context only the two most important ones here. MES for the process industry and MES for discrete industry. In the first case, machine and plant control systems will form a very great part of MES. In the second case, MES is more an online information system, a feedback and control system for production. Of the attempts to achieve standardization which we have mentioned, only a few need to be discussed here:

MESA

MESA (Manufacturing Execution Solutions Association) already has the concept as part of its name and as the first organization to adopt this concept is probably the most experienced to report on it. MESA's approach here is a very pragmatic one and describes twelve function groups which are required for an effective support of production management. These function groups are:

- Detailed planning (operation/detailed scheduling)
  Sequence and time optimization of the orders finely tuned to the performance of the machines including their finite capacity and to other resources.
- Resource management (resource allocation and status)
  Management and monitoring of resources, such as machines, tools, and so on.
- Registration and display of the current status of resources
- Document management (document control)
  Management and distribution of product, process, design or order information as well as work instructions which help secure quality.
- Material management (dispatching production units)
  Management of the input materials and intermediate products used in production, this in some cases being for the purpose of documenting material consumers.
- Performance analysis
  Comparison and evaluation of measured and recorded actual values for installations or areas against operational targets, customer targets, etc.
- Order management (labor management)
  Control and definition of operations and dispatching to work centers and personnel.
- Maintenance management such as maintenance and servicing
  Planning and implementation of suitable measures aimed at enabling machines and installations to meet their performance targets.
- Process management
  Control and management of the workflow in a production facility in accordance with the planned and current loads and specifications.
- Quality management
  Recording, tracking and analysis of the product and process, and verification against ideal values.
- Data collection and acquisition
  Visualization, recording, collection and organization of process data, of material and raw materials, of personnel handling, of machine functions and their control.
- Product tracking and genealogy
  Documentation of all events connected with the creation of a product. Recording details of the input materials and ambient conditions.

All of these function groups or reasonable combinations of them can form a total MES solution. At the 2004 MESA conference held near Chicago the term C-MES (collaborative MES) was coined. MESA's own MES functions as described above were redesigned and in some cases merged or given modified meanings. Here MES not only serves as an intermediary

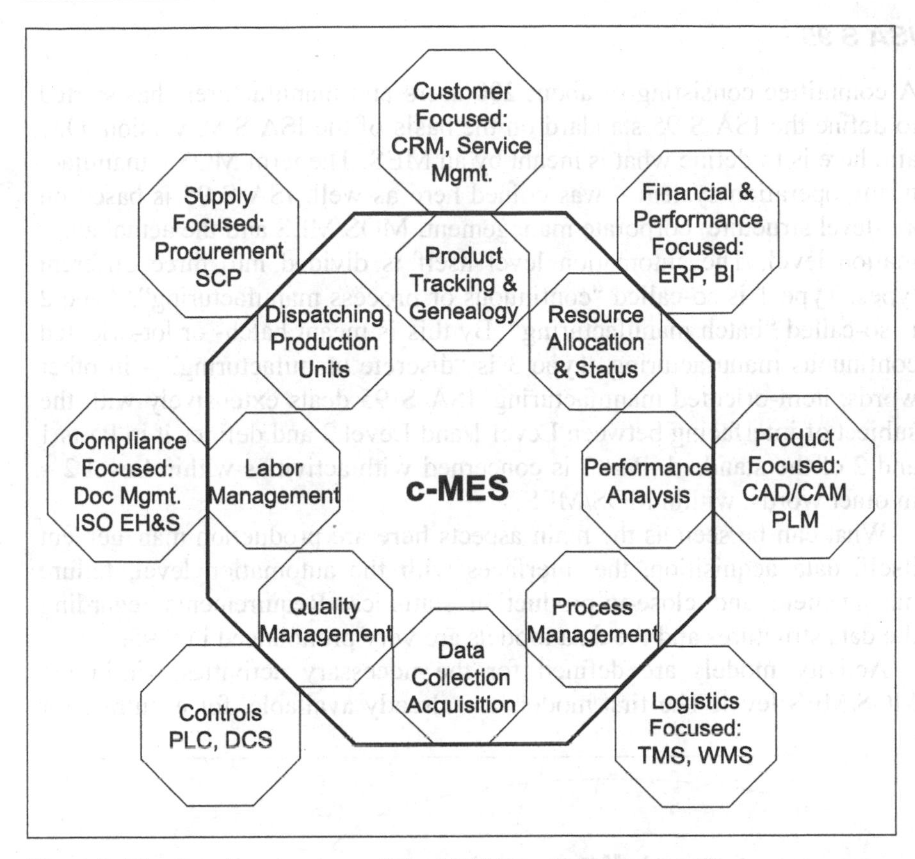

Fig. 1.7. The function structure of the MESA model

between automation and corporate management but it is also seen as a data and information turntable or hub. In this instance MESA goes so far as to regard C-MES as the integration platform within a company. According to this approach, even business, technical and logistical functionalities are just users accessing the MES platform. No doubt this is a very broad interpretation. In a manufacturing company, production is however the central key element and it must therefore be regarded as not entirely absurd to group all services around this important aspect. MES is here seen not only as a collection of functions in production management but also as an integrational hub for information throughout the company. At the 2004 MESA conference MES was also presented as an important instrument in the competitive struggle. According to this MES is supposed to give its users the advantages they need to be able to survive in international competition.

ISA S 95

A committee consisting of about 200 users and manufacturers has started to define the ISA S 95 standard on the basis of the ISA S 88 version. One aim here is to define what is meant by an MES. The term MOS – manufacturing operation system – was coined here as well. ISA S 95 is based on a 3-level structure: corporate management, MOS/MES and the actual automation level. The automation level itself is divided into three different types. Type 1 is so-called “continuous or process manufacturing”. Type 2 is so-called “batch manufacturing”. By this is meant batch- or lot-oriented continuous manufacturing. Type 3 is “discrete manufacturing” – in other words, item-oriented manufacturing. ISA S 95 deals extensively with the subject of interfacing between Level 1 and Level 2 and defines it in Parts 1 and 2 of the standard. Part 3 is concerned with activities within Level 2 – in other words, within MOS/MES.

What can be seen as the main aspects here are production management itself, data acquisition, the interfaces with the automation level, failure management and close-to-production statistics. Requirements regarding the data structures and the data models are very pronounced in S 95.

Activity models are defined for the necessary activities within the MOS/MES level. The first models are already available, for example, for

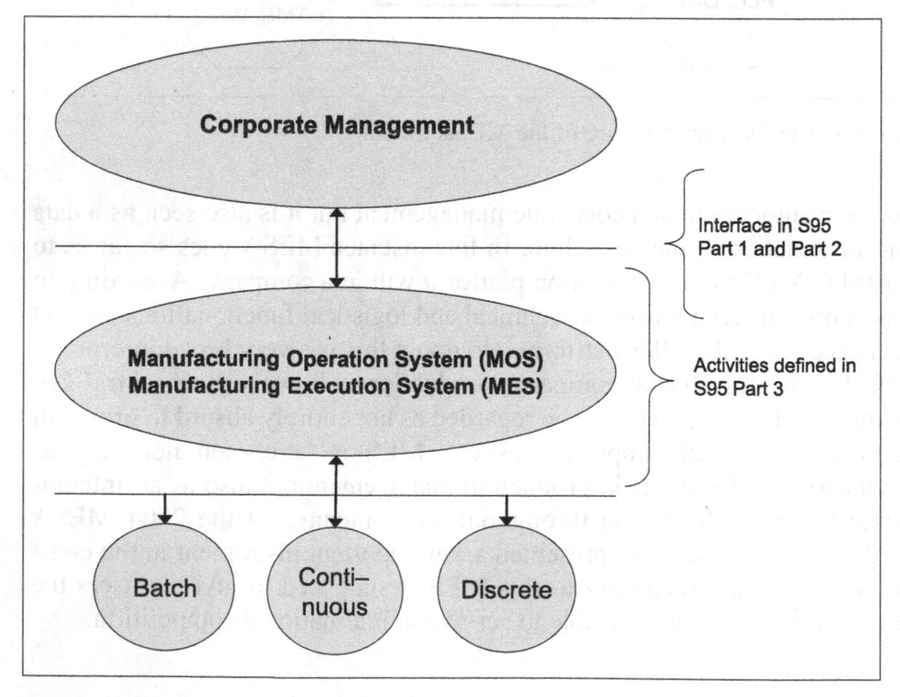

Fig. 1.8. The ISA 3-layer architecture

plant maintenance operations, quality assurance operations and for the production activities themselves. A very clear model structure is used for these activity models as well. Due to this very principle-based approach to MES, S 95 is more suitable for large systems which in many instances – for example, within a concern – have a certain degree of standardization. In daily project work, the application of the quite theoretical principles of ISA S 95 is more of a burden. Notwithstanding this, the S 95 concept is suitable for use as a design guide for some basic structures. ISA here describes a 3-layer architecture and introduces not only the term MES but also the term manufacturing operation system (MOS) . The manufacturing methods distinguished are discrete manufacturing, continuous manufacturing and so-called batch manufacturing.

NAMUR

NAMUR is a group of end users particularly involved with systems in the process industry (chemical or pharmaceutical industry). NAMUR's recommendations are based on the ISA S 95 definition and the group goes beyond the standard itself to make more concrete definitions. Functions and information flows are described in the form they take in the process

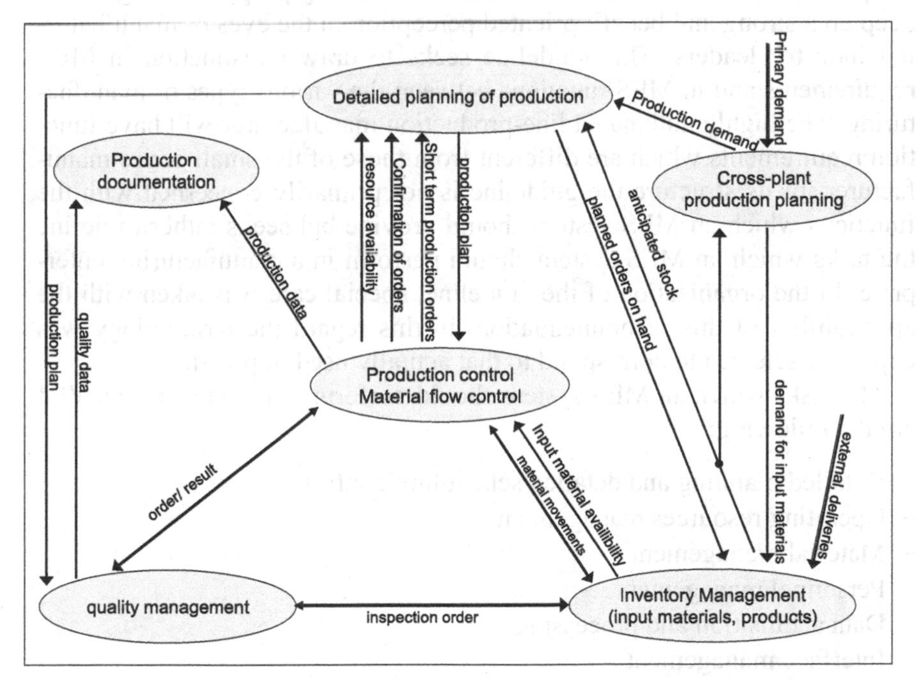

Fig. 1.9. In its recommendations the NAMUR organization provides a very detailed overview of the functionalities best implemented in the process industry

industry in particular. Even the line separating the actual automation level and MES is not so clearly marked here. The probable reason for this is that in the process industry important parts of production management have to be mapped into the machine or plant control systems themselves before quality or effectiveness targets can be reached.

In the case of a few large installations within a plant, even the boundary between rough planning and detailed planning, or medium-term and short-term planning, is not to be drawn as distinctly as in a shop-oriented production environment where there is a disproportionately greater possibility of product variants in production. The NAMUR recommendations thus represent good practical information for setting up MES systems in the process-oriented environment.

VDI

In 2004 the VDI (Verein Deutsche Ingenieure), on the basis of the above-mentioned standards and of current findings and market developments, started work on defining the term MES in a VDI guideline . A special concern of this guideline is to give the term MES a fixed meaning and thus prevent it from being misused by certain system vendors, who tend to jump on every new "buzz word" for pure marketing purposes. Target is to keep up a strong and benefit oriented perception in the eyes of manufacturing industry leaders. This guideline seeks to draw a distinction in MES requirements and in MES functions between the various types of manufacturing. The highly automated line-production manufacturer will have function requirements which are different from those of the small-series manufacturer. In its structure the guideline is not primarily concerned with the functions which an MES system should provide but seeks rather to define the tasks which an MES system should perform in a manufacturing enterprise. In the organization of the guideline, special care was taken with the applicability of this recommendation. In this regard the terminology was expressly selected to correspond to that actually used in practice.

The tasks which an MES system should perform in the view of the VDI are the following:

- Detailed planning and detailed scheduling control
- Operating resources management
- Material management
- Personnel management
- Data acquisition and processing
- Interface management
- Performance analysis

- Quality management
- Information management

1.4.3 The ideal MES

When we consider the tasks we have so far described, tasks which an MES should perform within the context of different control strategies and different kinds of manufacturing types, the question arises as to what the ideal MES is. Is there an ideal MES anyway?

Throughout all manufacturing industries, "yes" will certainly not be the answer to this question. Nevertheless an attempt should still be made to sketch out the functionalities an MES should ideally have for the environment of discrete manufacturing or even of batch-oriented manufacturing. The "working area" of an MES will, of course, extend from interfacing with corporate management applications as far as the deepest depths of data acquisition, communication with industrial systems and the provision of data for machine controllers or influencing the machine controllers directly. This extensive operational area, which not only includes a comprehensive range of different topics but which must also cover different time levels, ranging from days and weeks down to seconds, also calls for a tiered examination of the individual functionalities. This examination can thus be subdivided into the following areas:

- The functionalities of an MES itself
- Communication with corporate management applications
- Communication with the manufacturing environment

As a rule a PPS covers three function groups: production, quality and personnel. Within these function groups powerful modules are available which can be activated and used depending on requirements. A basic system ensures that all modules are linked together in real time and furthermore provides intermodular functionalities. Another important function is acting as the information hub.

In an analogous manner to these function groups in the PPS/ERP system, an MES can also be divided into three function groups. These are primarily the functionalities for production, the functionalities for quality and the functionalities for personnel allocation. To prevent these functions from appearing rather too abstract, we shall use the descriptions and in some cases the designations used in the classic definitions of the modules from which the MES developed.

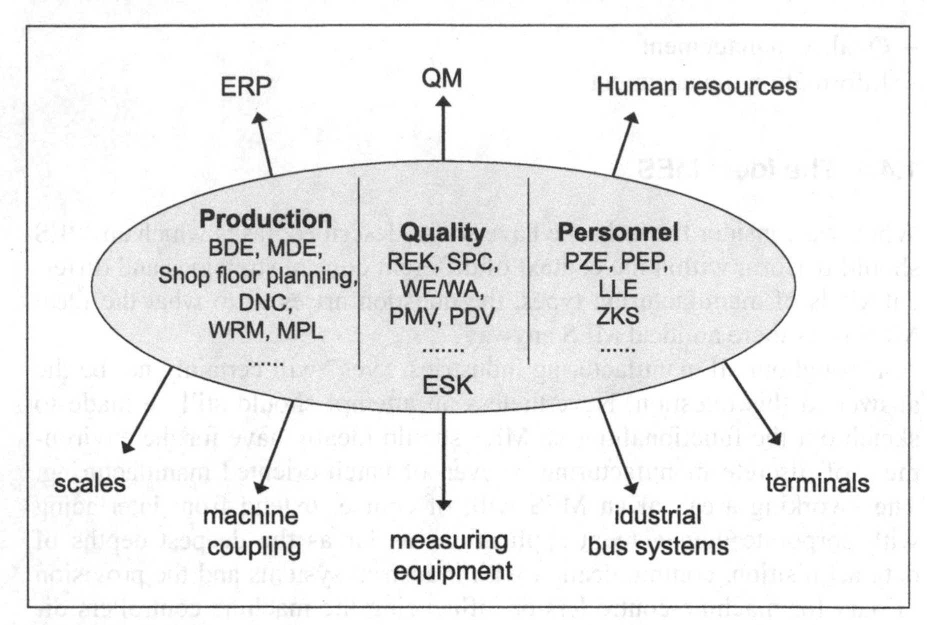

Fig. 1.10. MES function groups

Function group: production

The function group "production" can include the following modules:

- PDA: production data acquisition
  Here order- and person-related times and quantities are recorded. In the quantities a distinction is drawn between conforming items and scrap and also scrap categories. It is also possible to directly record material consumption and wear and tear on operating resources or process materials and associate them with the corresponding orders. The data which have accumulated over shifts, days or weeks are prepared correspondingly and then made available to corporate management applications. Detailed, real-time displays and analyses can be prepared in parallel with this for organizational functions in production.
- MDC: machine data collection
  Machines or other operational resources are managed in this group. Status data can be acquired manually and automatically via comprehensive systematics and subsequently be assigned to resources or resource groups. The data can then be supplied here not only by conventional terminals but also by industrial bus systems. Automated registration of quantities via counters, via balances, scales and comparable equipment should also be supported. The data so acquired can then be supplied in a condensed form to corporate management as a basis for effectiveness

statements but also, in their detailed form, permit analysis of points of weakness within production.

- Control station, planning table
  These functionalities are the subject of vigorous discussion. On one hand, ERP systems via their APS functions offer planning possibilities within a shift. In the case of the control station and planning table modules in an MES system the emphasis is even more on preparing technologically feasible plans whereby the feasibility should also be assessed on the basis of a current situation. It is a waste of time making plans whose likelihood of being put into action is relatively low. From this it follows that the more detail and precision are required for planning, the more the plan must be related to current situations. Here the detailed planning modules of an MES should not only allow manual interventions which are simple from the operator's point of view but also offer support for fully-automated loading as well as simulation and optimization.
- TRM, DNC (tool and resource management and transmission of machine settings)
  An MES should be able to manage tools and other auxiliary materials and equipment on a technically oriented basis. It is less a matter here of managing inventory (as is necessary and normal in the field of corporate management) but much more a matter of the technical state of operating resources, of current availabilities, of managing compatibilities with machines and the qualitative evaluation of these auxiliary resources. From the direct proximity to machine data collection, possibilities also emerge for preventive maintenance and thus for being an effective means of being able to reduce the occurrence of unforeseen downtimes.
- MPL: material and production logistics
  A particularly important point in production is those materials which are in circulation or being held in interim storage facilities. Here a material and production logistics (MPL) module helps to keep an overview and to initiate transportation activities at the right time. A function of this kind should not be confused with a warehouse management function. Instead it is exclusively concerned with WIP (work in progress) – in other words, the materials which are circulating outside the classic storage facility. MPL supplies information about the quantities which are currently in circulation.

Function group: quality

The quality function group in the sense of an operative quality assurance should not be confused with the kind of quality management offered by, for example, the major ERP systems and which is to be understood as a company-wide quality management system which plans and administers.

- SPC: statistical process control. SPC is concerned with connecting up measuring equipment to allow the specific acquisition of measured values, online comparison of the measured values against a set-point value, and the immediate output of warnings should the two values differ by earlier specified levels of tolerance. Naturally, SPC saves these random samples and permits tracking of certain trends. These trends can also be displayed directly online in the production facility and thus identify and prevent any production which might have malfunctions.
- NCM: non-conformance management. Here products which are the subject of complaints are traced back on the basis of technical aspects, manufacturing conditions, and input materials. Countermeasures are introduced and tracked via a response measures system.
- Incoming goods. In this context, the incoming and outgoing goods function is less the rating of suppliers as is normal in QM but rather the specific registration of delivered goods or of dispatched goods, the verification of batch numbering, and an online alarm facility if certain values are out of tolerance.
- Inspection equipment management is directly comparable with TRM from the function group "production". Here inspection, measuring and test equipment is managed and it is ensured that they meet the required standards and that they can also be used for the corresponding tests and inspections. This information must, of course, be available with immediate access during the inspection and testing procedures.
- PDP. An unusual feature in this context is the process data processing module. This implements the proposition that quality is not simply a property of the product but that quality also depends on the circumstances of production. For example, the pressures and temperatures used in the production process are of decisive importance to quality and for this reason an MES should also have the capability of acquiring process values directly, verifying them against tolerance or intervention limits, and in the event of errors recommending countermeasures.

Function group: Human Resources

- Staff work time logging. Throughout its history the human resources function group has always been very closely attached to corporate management. A series of elegant simplifications emerges for personnel allocation and personnel management in production when production-related personnel handling is mapped into the MES system. The staff work time logging module covers clocking-in/clocking-out data as well as absence times and can also keep month-based time accounts where this would be useful. This approach would be particularly ideal when labor capacities in production facilities play an important role (as is very often the case) and these labor capacities have to be scheduled in a timely and efficient manner.

- Incentive wages. Staff work time logging in the MES deploys its strengths when bonus systems need to be implemented. Due to its direct proximity to production data acquisition, the incentive wages calculation module can very effectively create the connection between absence times and work order times and thus greatly simplify calculation of performance levels.
- Short-term manpower planning. With the short-term manpower planning module it is possible, in a similar manner to the planning table in the production function group, to obtain an overview of active personnel and to prepare the corresponding personnel schedules in an elegant manner or even with the aid of automatic functions, these schedules taking into account the loading situations on a departmental, company or plant level. Once again it is true to say that plans to be implemented at short notice can only be effectively drawn up on the basis of an up-to-date image of production.
- Access control. If staff work time logging has been integrated into the MES, one simple side-effect of this is that access control in the production facility can also be implemented via the access control module.
- Escalation management. Other mechanisms can be incorporated in the overall concept of an MES to make prompt reaction possible and, should limit values (what is meant here in the broadest sense is qualities, utilization ratios, downtimes, and so on) be violated, to make it possible to automatically implement the corresponding escalations or alarms and thus bring about a considerable reduction in how long faulty operational states last. An escalation management capability provides support here for production controlling and the operative level.

1.4.4 Technical requirements

Data storage

In an MES, data storage should always be right up to date technically and use standardized databases. One especially important point is the adaptability of an MES. Production systems are structured in various ways. Processes in different production systems cover a very wide range. A standard solution can scarcely be powerful enough for it to be possible to make all conceivable modifications via parameters. In addition, the complexity of a parameterization must be readily comprehensible. The happy medium here is parameterization which can still be carried out by users coupled with the possibility of controlling processing instances with the aid of suitable resources and developing one's own applications. Here the modern user can expect an MES system to include a solution designer which allows him to develop his own applications on different levels and to incorporate them into the menu system. Straightforward command lines should also enable the user to influence processing instances in the MES so as to obtain *his* results as effectively as possible. As far as displaying information is concerned, well-parameterized analyses should, of course, be seamlessly available. These analyses should also be configurable within a reasonable framework on a user-specific basis.

Communication with corporate management

An MES must have the appropriate interfaces which as a standard feature can communicate with the most common ERP, personnel and QM systems on the market. In all cases where the products concerned tend to be atypical, an MES should provide easily parametrizable interfaces which can be adjusted easily.

Communication with production management

The present-day manufacturing environment does, of course, offer a wide range of production facilities from which an MES must not only tap off data but conversely also supply data in order to influence this environment. Connections with machines and weighing devices are used for acquiring data regarding quantities, qualities and even points of weakness. Libraries are maintained here which allow a straightforward connection even to non-standardized products. More and more frequently, commercially available machine and machining centers are connected via their own industrial bus systems. An MES has the communication modules needed for reading the desired data from these bus systems. Important buzz words here – which are also dealt with in the chapters which follow – are OPC and, for example, Euromap 63, as also certain manufacturer-specific systems.

Data collection terminals play a very important role in MES systems. In the past they were plain data collection devices but today they are more and more becoming information media as well. For this reason a powerful MES should support not only simple input devices but also PC-based input and information stations. In the case of the more complex terminals and PC-based systems, today's user can expect a user-friendly information and data collection interface as well as plausibility checking which signals faulty states or erroneous inputs immediately upon input. Here PC functionality is used to transmit directly by electronic means everything which is still to a great extent circulated in paper form through the production facility.

The function groups and individual modules shown here should be taken as examples but today do nevertheless represent a major part of what a powerful MES system should offer the user. At this point an important issue still needs to be made: an MES should be structured on a modular basis so that it can be introduced gradually as required and progress in familiarization with its functionalities is made easier.

1.5 Vertical and horizontal integration

In the past a two-layer model arose automatically from the two levels of automation and ERP. In most cases information was exchanged manually between the two layers.

The connection between the levels of corporate management and of production was thus a very indirect one and communication cycles were designed with a resolution of several days. With the development of the MES concept, even production management, as seen from the IT side, became a distinct discipline. It has now become possible to assign really specific fields of action to the three levels of corporate management, production management and production. In its dealings with ERP or PPS, corporate managements tend to work on a long-term basis with a resolution of weeks or months. Rough production planning deals with a medium-term range of weeks or days, while detailed planning – also known as load planning – acts in the short term in days and shifts. Decisions taken within production management must be made within a time range extending from shifts down to minutes while automation with its machine and plant control systems needs, of course, to react within minutes or even seconds.

The corresponding diagram symbolizes a control characteristic within the different levels whereby the control cycles run within the time resolutions shown. There are, of course, no precise boundaries between the three levels of a manufacturing company. Accordingly, between ERP and MES we find the APS (advanced planning and scheduling) functions which, depending on the type of production, tend to be closer to ERP or to MES.

Just as fuzzy is the demarcation between the MES and automation. The presence of functions such as data acquisition and the transmission of machine settings alone creates a tight connection between the two levels which should still nevertheless be regarded as separate levels and whose character tends to be related to planning on the one hand and to technical implementation on the other.

In the first sections of this chapter three different types of production are examined. Figure 1.11 seeks to answer qualitatively the question "How much does each production type need MES?"

As we have already said, the functionalities of APS and of system control are located in the overlap areas of the three corporate levels. For the sake of clarity we can also say that in the demarcation between ERP and MES functionalities, MES tends to act with a technological orientation and in real time while activities in ERP tend to be commercial and cut to the medium term.

Not only that but the three MES levels of a manufacturing company work with completely different time horizons. Within these time horizons they are autonomous control systems whose control cycle is oriented by the corresponding horizon.

The definition of a discrete type of production indicates that in the processes there are many degrees of freedoms with the aid of which a production order can be channeled through production. Of course, degrees of freedom also mean many opportunities for long transfer times, waiting times and inefficient order sequences.

Situation- and technology-oriented planning aids are required here which enable production management to react swiftly and appropriately to errors. An MES covers part of the APS functionalities at this point. At the border with automation the MES permits direct data tapping within production.

The situation is somewhat different with mass production and assembly line production. Here the difference between rough and detailed planning

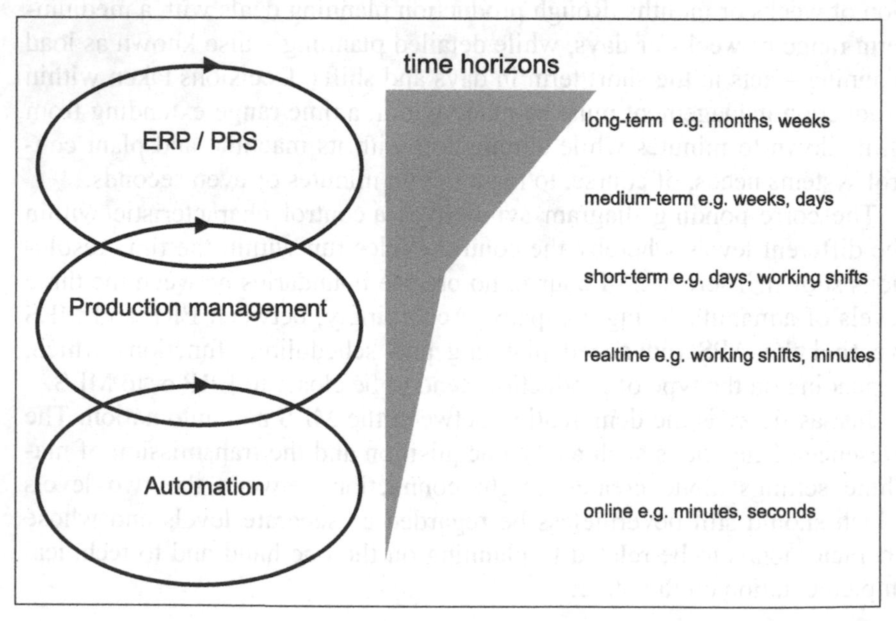

Fig. 1.11. Requirements for MES penetration within the company

is less than in discrete manufacturing. Relatively long-running orders and lengthy retooling times rule out short-term reactions and thus automatically make medium-term production schedules necessary. A major part of short-term activities is handled on the automation level – in other words, in the machine and plant control systems. With this type of production the automation level is very powerfully developed. This reduces the planning portion of the MES to a relatively narrow field which is limited to a short-term and technology-oriented situation presentation and to evaluations relating to times, faults and qualities.

For the make-to-order manufacturer the resolution of large bills of material is a theme of particular interest. His business is governed by long-running orders. In such a case, a short-term, reactive detailed planning capability will only be of any use in make-to-stock production areas. A major part of automation is also found in these areas, apart from, of course, complex machining centers in which extensive machining activities are carried out on individual parts. Here the MES part will be limited to the field of small-batch products and to time recording and project time recording inside the plant engineering and construction operation itself.

Different types of production need different MES functionalities and even need MES implementations of different levels of capability. Figure 1.12 shows the extent to which MES functions can be covered by ERP or automation or can extend into these areas.

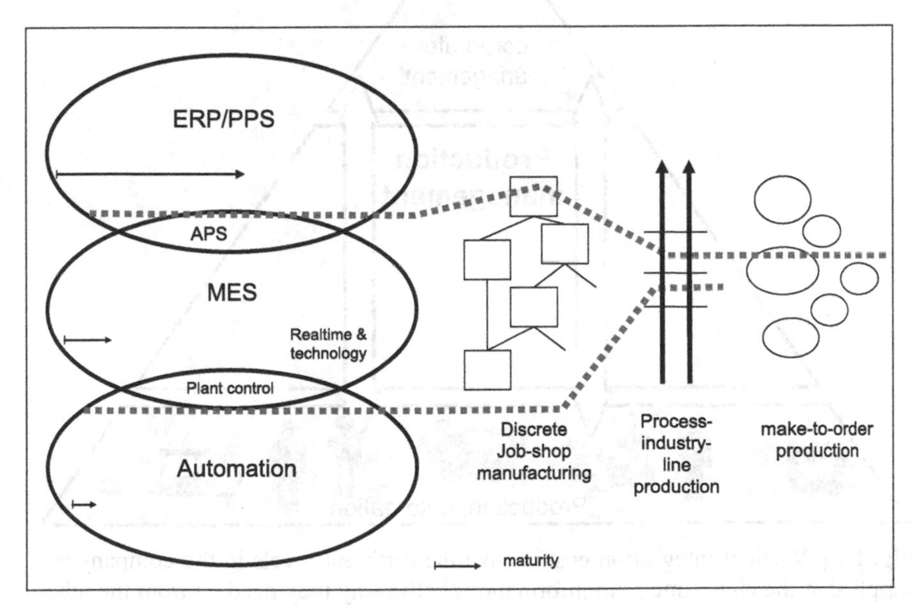

Fig. 1.12. MES requirements by type of manufacturing

With the 3-level structure we have described it has been possible, using the MES as a linking element between corporate management and production, to achieve vertical integration throughout. In the IT world, with a model of this kind there is now no necessity for any instances of modal fragmentation in which protracted and time-shifted manual recording and data acquisition routines prevent information exchange in real time. Via the MES the ERP level can now supply production with up-to-date information in real time. Via the MES, ERP receives back correctly prepared information at the right time. However, at this point the MES does supply production management in real time with the technology- and situation-oriented information which is needed for a timely response to faults and malfunctions or to keep fault conditions as short-lasting as possible. Despite the tight coupling afforded by vertical integration the three time levels are decoupled to the extent that each level can act correctly and the company as a whole can still map time domains ranging from long-term to online.

The attention currently being given to vertical integration shows how important the role of the MES is in the architecture of a company. According to what has already been said, within production management – in other words, in the domain of the MES – there are three function groups: production,

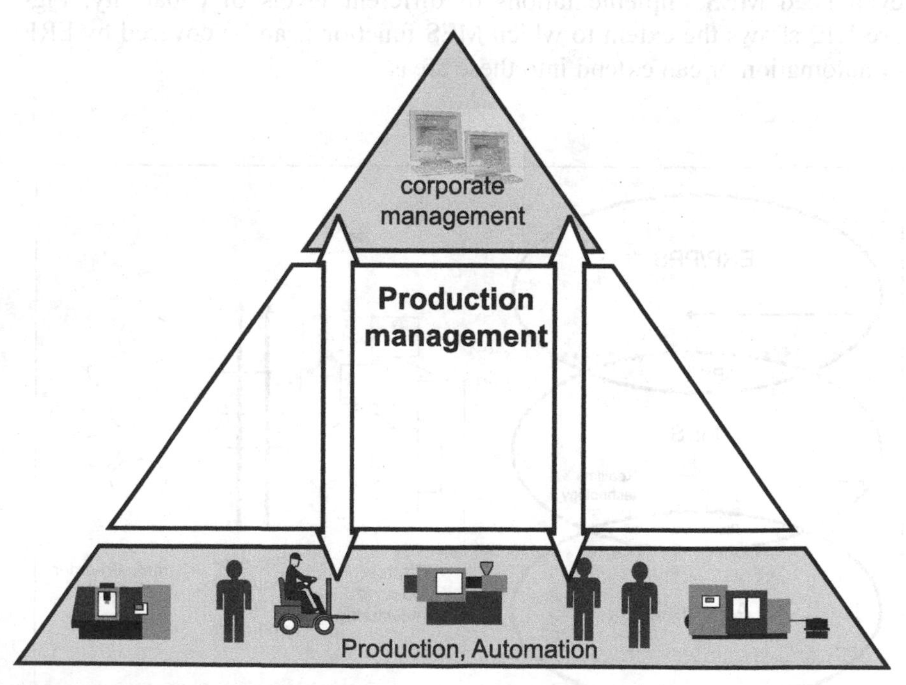

Fig. 1.13. Vertical integration ensures that the different levels in the company are supplied at the right time with information (in the way they need it) from the other levels

personnel and quality. In very few cases indeed are these function groups independent of each other. To manufacture something, the right personnel are needed at the right time and this expensive resource, the individual, should be employed as effectively as possible at the right place. In production, parts can be produced with different qualities and these need to be inspected. This short list alone demonstrates that production management, if it is to function effectively, must attend to all three function groups more or less simultaneously. From this arises the demand that these three function groups need to be connected very closely together. According to this, the IT system which maps these functionalities should act in the most uniform manner possible and be based on a single data pool. This prevents duplicated acquisition of data, redundant master data and transaction data.

This so-called horizontal integration is thus an important requirement whereby an MES system provides production management with effective support. The above considerations show how important this integration mechanism (the MES system) is to vertical integration in a manufacturing organization. Vertical integration is an important requirement if a manufacturer is to enjoy a competitive future.

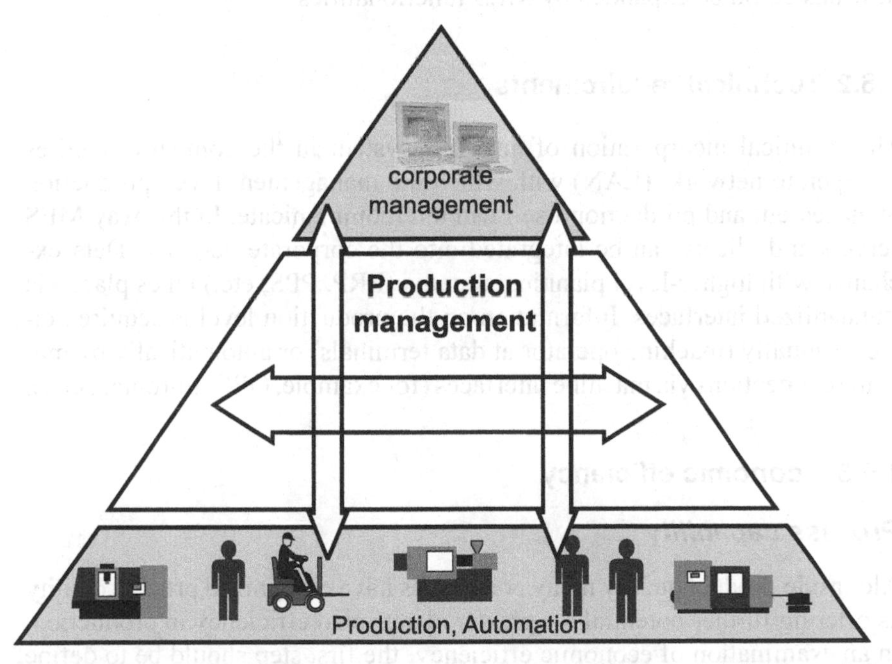

Fig. 1.14. Vertical integration is important for the effective functioning of the entire company. Horizontal integration in the MES domain is a special instance of implementation and permits effective working with the different function groups in an MES

1.6 Use of an MES system in the company

1.6.1 Organizational requirements

The use of an MES in the company depends less on the size of the company or the particular industry but rather on the production structure (shop production, production segments, assembly line/flow production, etc.) and the type of production (non-repetitive production, one-off and small-series production, quantity production, mass production). While the main reason for using an MES in the case of multistage shop production (turning shop, milling shop, electroplating shop, etc.) is to improve the interplay between the individual processing steps and thus the passage of the order through production, the emphasis in the case of mass production is certainly on increasing the degree of utilization of individual production lines. Due to its modular structure, MES can be easily adapted to the specific production environment and its task demands. Within the context of a planned MES project it is therefore important to first ascertain what initial situation (production structure and type) applies. The second step should be to find out how production planning and control is currently being carried out and how this could be expanded by MES functionalities.

1.6.2 Technical requirements

The technical incorporation of an MES system in the company requires a corporate network (LAN) with which the management level, production management and production itself can intercommunicate. In this way MES servers and clients can be integrated into the corporate network. Data exchange with higher-level planning systems (ERP, PPS, etc.) takes place via standardized interfaces. Information on the production level is acquired either manually (machine operator at data terminals) or automatically by machine connections via machine interfaces (for example, OPC, Euromap, etc.).

1.6.3 Economic efficiency

Process capability

Alongside product quality many companies have recognized process quality as offering further potential for greater operational efficiency in production. In an examination of economic efficiency, the first step should be to define which process-oriented objectives are to be achieved by the use of an MES. Goals of this kind could be, for example, a reduction in the lead time, an increase in machine utilization, an improvement in delivery reliability, a reduction in work in progress inventories or a reduction in defect costs. On the basis of the objectives which have been defined it will be possible to investigate in concrete terms the potential for economic improvement. Examples of potentials of this kind include:

Increasing machine utilization

In the metal-working industry the average machine utilization is often lower than assumed but planning and calculations are actually based on a higher assumed utilization. Systematic recording and evaluation of all instances of unscheduled downtime with the aid of an MES system helps to reveal the causes of downtime, remove them, and thus considerably improve machine utilization. The investment costs of an MES here only amount to a few percent of annual machine costs. In other words, even an improvement in machine utilization of just a few percent will deliver the return-on-investment (ROI) desired.

Reducing the lead time

Reduction in lead times is without doubt the most important factor in economic efficiency in production. Connected with the order lead time are delivery time (competitive advantage), delivery reliability (customer satisfaction), inventory (liquidity) and throughput (profits). Using an MES system makes it possible to recognize these potentials and exploit them systematically.

These simple examples show what process potentials can be hidden in a company and how they can be uncovered with the pragmatically applied software tool MES alone. Generally speaking, potential for improvement in the company can be picked up more quickly due to the MES-based process transparency and also exploited more quickly due to the MES-based control cycles. Furthermore, since all process data are acquired systematically, it is possible to detect more potential for improvement than without an MES.

1.6.4 Support for CIP and current certifications

Although there is still huge potential hidden in the improvement of process quality and although the continuous improvement process (CIP) is anchored in the current certification standards, such as DIN EN ISO 9001:2000 or ISO/TS 16949:2002, in many companies process orientation is not actually implemented in practice. A frequent reason for this is a lack of process transparency, although this situation can easily be remedied by using an

MES. Due to the integration between the ERP system on the one hand and the production level on the other, an MES is continuously acquiring data for most process influences (orders, machines, tools, personnel, material, quality, etc.) in production. This means that hit lists (Pareto diagrams) relating to the most frequent causes of problems and errors can be drawn up, process times determined (setup times and processing times, waiting times and downtimes, interruptions due to faults, etc.) and key data relating to process and product quality calculated and displayed (for example, the OEE index, machine utilization, degree of processing, scrap rate, etc.). In this way an MES system supports improvement activities in all phases: define (definition of processes to be improved), measure (measure process data), analyze (analysis of measured data) and control (check steps taken).

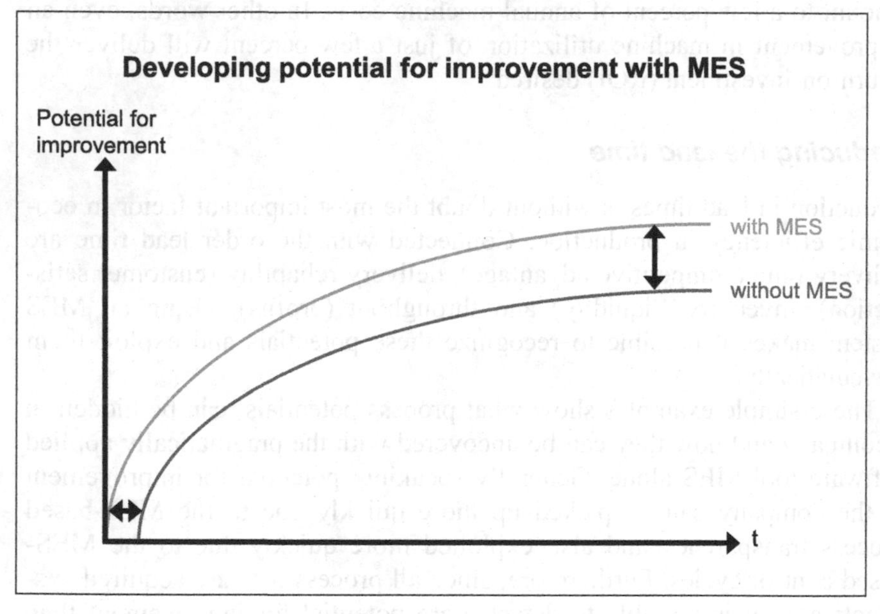

Fig. 1.15. Faster detection and exploitation of improvement potential with the aid of an MES

1.6.5 Definition and tracking of objectives

Manufacturing scorecard: process-oriented key data

To be able to even survive in an ever fiercer competitive environment, many companies depend on an entirely specific competitive strategy, such as price leadership, technological leadership, service level, flexibility and so on. With the aid of the manufacturing scorecard method and proceeding from this strategy, process-oriented key data for production can be derived on the basis of the measured variables supplied by the MES and can then be communicated to employees as targets. The effect of this is that the employees in production act in harmony with the interests of the company and will also ponder on how their own key data could be improved. Even CIP activities are supported by these key data since suggestions are predominantly made on the basis of key data and are then directly reflected in the target figure. Thus knowing the current status and also the target figures is an immense motivation for the employees. At this point, in order to demonstrate that MES can offer new possibilities even with familiar motivation mechanisms, let us mention two examples which have long dominated the manufacturing environment.

Group work

Since a process is in most cases affected by several people (for example, the setting-up process by the machine operator, by the line engineer and by the toolmaker) it is apparent that the same key data can be supplied to precisely those employees (the group) who can jointly influence the result of the process. An MES supports this group work by providing all relevant information in paperless form on so-called group i-point terminals.

Target agreements and bonus payments

The key data on the manufacturing scorecard are suitable not only for operational target agreements but also for a bonus payments scheme, something which gives a further boost to employee motivation. Examples of these key data are the degree of utilization = main usage time/occupancy time; the degree of processing = main usage time/lead time; capacity used = occupancy time/lead time; the labor utilization rate = main usage time/attendance time; and the OEE index = availability × performance × quality.

1.7 Practical examples of potential benefits

When an MES system is applied rigorously within a manufacturing organization, the presence of control cycles with short cycle times and the intensive interconnection of several classic disciplines such as control station, PDA, quality assurance, machine data collection, and so on, delivers a series of special potential benefits which can be listed here in a summarized form.

- Short-term scheduling takes capacity limitations into account and ensures a delivery date based on an up-to-date model of the capacity situation. This in turn makes possible a considerable improvement in on-time delivery performance – in other words, customer satisfaction and employee motivation. With a simulation based on current production states the best possible alternatives can be selected with the aid of scenario creation. This can deliver the following advantages: better on-time delivery performance and improved capacity utilization while lead times and inventory levels can be reduced. Similar effects are generated by an up-to-date planning table which deploys ergonomically-presented order information to increase the scope of action of the planner or operations planner and scheduler.
- The coupling of PDA, MDC and planning table means that very realistic, real-time images – in some cases even using automatic counter data acquisition – can be included in a planning table and the above-mentioned effects will also result.
- Technology-oriented order and article statistics from PDA, MDC and quality assurance can bring to light eventual technical circumstances which result in higher costs due to unplanned consumption of material and/or time in the manufacturing process itself.
- Real-time order, machine and personnel overviews improve the quality of information given to customers regarding delivery dates and can also reduce undesirable in-process inventory levels.
- A higher level of data consistency and less outlay on data collection is achieved by means of an intensive coupling of CAQ and PDA. Above all, immediate statements are possible about how many parts are currently in stock and in what quality and where direct action is required regarding secondary finishing or quality problems.
- When tool or resource management is coupled with machine data collection the result may be a reduction in tool or machine failures. Detailed information relating to operating resources is available online and maintenance can be scheduled for the appropriate times.
- Downtime evaluations and weakness analyses from machine data collection will result in higher utilization ratios and thus more capacity for the same cost.
- Interlinking staff work time logging or time management and production data acquisition can make calculation of performance-related bonuses a simple matter. In this way company objectives such as machine utilization, on-time delivery performance and quality can be given direct support. With these mechanisms it is also possible to log times by their causes, thereby permitting overhead costs to be lowered throughout the company.

- Inventory overviews obtained by connecting together materials management and PDA may allow considerable reductions in inventory levels of production materials in material buffers and interim storage facilities and thereby reduce capital lockup as well.
- Coupling together process data processing, CAQ and MDC means that seamless product documentation can be maintained. This is standard in regulated industries and more and more demanded by automotive suppliers and food industry. It means that violations of tolerance and intervention limits can be documented and fast reactions to negative trends are also made possible. Displaying production-relevant documents on a data collection terminal (the documents may come from all possible operational divisions) permits a considerable reduction in outlay related to the information accompanying production. With this mechanism, a form of production which involves little paperwork becomes ever more realistic.
- With batch and lot tracking made up from CAQ, PDA and materials management it is possible, as has already been stated, to maintain seamless product and production verification documents. This does however also make it very easy to provide traceability information for input materials and scrap. By that it becomes possible to uncover improvement potentials in production.
- By condensing the immense amounts of detailed information held in an MES it is possible to generate management information which reflects the technical situation and which also enables conclusions to be drawn regarding potentials for improvement, such as, for example, the OEE index for individual operating resources, entire departments or even entire plants.

2 MES for process capability

2.1 Economic efficiency as a process property

Today the economic efficiency of the company is scarcely a property of the products any more but rather of the processes. Companies today are thus faced with the task of optimizing their process chains, something which in practice leads to a reorientation in resource steering which is decisively important in competition: while in the past attempts were made to exert control over the economic efficiency of production on the basis of figures from the company's accounting department, the approach today is to try to identify the processes lying behind these figures. That, in brief, is how Norton and Kaplan tackle the problem with their "balanced scorecard" (Kaplan and Norton 1997). The method which has been widespread to date is to organize resources by the result via costs. This approach is foundering on account of the increasing proportion of overhead costs since it forces the cost accountant to take a considerable cost block which for logical reasons cannot be allocated to the cost unit (products) in a manner corresponding to the cause of the costs and notwithstanding this to still allocate it to artificially created codes (cost center account) in a way unconnected with the cause of the costs.

The main points of criticism may be summarized as follows:

- Overheads are proportional to time – time consumption figures (lead times) thus become an extremely important cause of costs. They are not however picked up by traditional cost accounting – there is no time dimension – which in practice means that an instance of production with a long lead time including subsequent storage is calculated in virtually the same way as production with a short lead time.
- This in turn results in the considerable effort being expended on improving the efficiency of value generation (degree of utilization) not being expressed in the costs.

Opening up the areas of potential hidden in the processes is increasingly becoming a matter of sheer survival: against a background of intensifying competition virtually all companies are operating today with their backs to the wall as regards not only their prices but also their equity capital. Process potentials are, on the other hand, comparatively enormous. Today this means the first step should be to identify the process chains and then to keep constantly improving them.

2.1.1 The process-oriented approach of ISO 9001/TS 16949

A process-oriented approach is the basis of process control. "For an organization to be able to function effectively it must detect, direct and steer large numbers of interconnected activities. An activity which uses resources and which is carried out in order to make possible the conversion of inputs into results can be regarded as a process." (ISO/TS 16949 2002).

The process-oriented approach of ISO/TS 16949 thus covers:

- Understanding and meeting requirements
- Process assessment from the point of view of value generation
- Achieving results as regards process performance
- Permanent process improvement on the basis of objective measurements

Competition in the global market is shifting increasingly from competition between products to competition between processes. Companies such as Dell, Amazon or Würth may serve as examples of how markets can be created not via products but via business processes.

The way value generation is oriented towards the customer has consequences for the control of internal processes: the customer does not judge isolated improvements in individual processing steps but rather solely the result at the end of the value chain – in other words, the capability of the entire process. This transition from the production economy to the service economy is today referred to as the second industrial paradigm.

Process capability brings a new way of looking at things: to steer the economic efficiency of the company not by means of its technical equipment but rather via its internal processes. While improvements on the basis of an improved manufacturing technology can only be achieved with difficulty – most companies have the very latest machines and tools and may well buy their materials from the same suppliers – the process potential is comparatively enormous. Although process-based certification standards have now become very common and although most companies have now become certificated on the basis of a process-oriented code (ISO 9001, TS 16949), the process-oriented approach is hardly observed in practice.

We shall therefore go on to show what action can be taken and tools deployed to achieve the process capability of the company in practice.

2.1.2 The process potential in figures

Process capability means the ability to work without errors. Process capability can be measured as a spread within preset specification limits. The process capability can be statistically expressed in numbers by sigma which is a measure of the scatter. Purity grades can be assigned to the different sigma values. Today purity grades are expressed as ppm values (parts per million non-conforming parts in a delivery).

Figure 2.1 contrasts the dependence between process variations – expressed as the scatter value sigma – and failures – expressed as the ppm value. The really important thing to realize here is that when the process capability is improved by just one sigma level, an improvement in economic efficiency can be achieved by reducing failures by one order of magnitude, something which is a long way from being achievable by making improvements in the processing (machines, tools, methods, etc.) (Rehbehn and Zafer 2003).

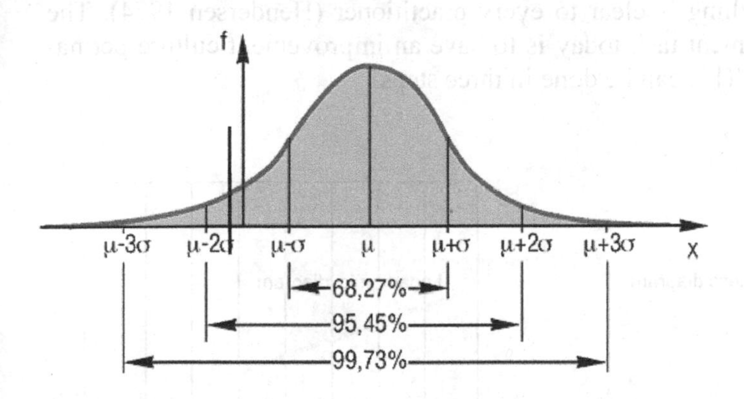

Sigma level	CPK values	Misperformance ppm	
2	0,67	22.750	
3	1.00	1.350	
4	1,33	32	factor 100
5	1.67	0,3	
6	2.00	0,001	

Fig. 2.1. Assignment of sigma level and failures

2.2 The process capability of the organization

An organization may be considered process-capable when it is capable of learning on a permanent basis. The learning potential of organizations (social, technical or economic) is in principle unlimited – but it must be systematically opened up. This presupposes that learning behavior is anchored in the organization. As in aviation, where every flying accident makes flying safer, in the factory every error must make the company better.

Figure 2.2 shows the potential an organization has for improvement by the example of the experience curve as developed by the Boston Consulting Group. This states that each time the cumulative production quantity is doubled, the unit costs will fall by a fixed percentage (20–30%).

The relationship between the costs and the quantity manufactured is based on the assumption that a company with rising production levels learns how to manufacture the products more cheaply – a relationship which for one thing is clear to every practitioner (Hendersen 1974). The critical management task today is to have an improvement culture permanently in place. This can be done in three steps.

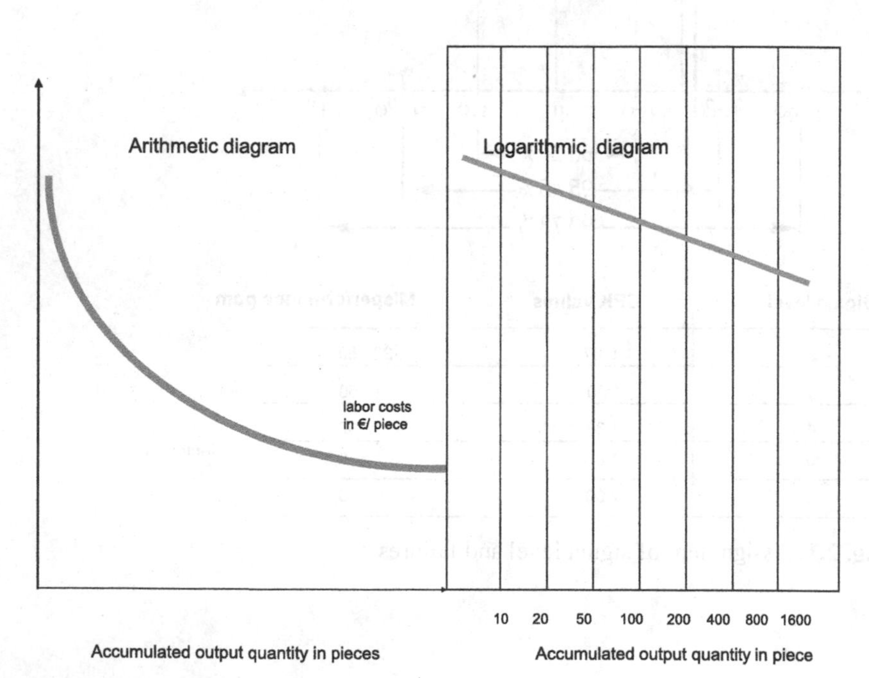

Fig. 2.2. The experience curve

2.2.1 Identification of systematic errors

Process control is not possible unless systematic errors are eliminated. Here the first step is to register and analyze all influences on the process. Figure 2.3 by way of example shows the errors which occurred for a machine, machine group, department or order and which can now be analyzed by the employees. Simultaneously with order log-on at the machine terminal, every MES (manufacturing execution system) registers with virtually no manual input the corresponding process parameters such as: machine, machine group, tool, article and operation, order, customer, shift, employees, and so on.

While random errors can be attributed to normal process variation and are thus difficult to overcome, it is of decisive importance to the analysis to identify the systematic component behind the process disruptions, since the establishment of a causal relationship is the essential condition for eliminating this error on a sustained basis. With regard to a particular cause, errors cannot be corrected until error clusters have been found. Error clusters can typically be recognized by statements such as "every time when …" (in the night shift, with this tool, with this customer, with this machine, and so on).

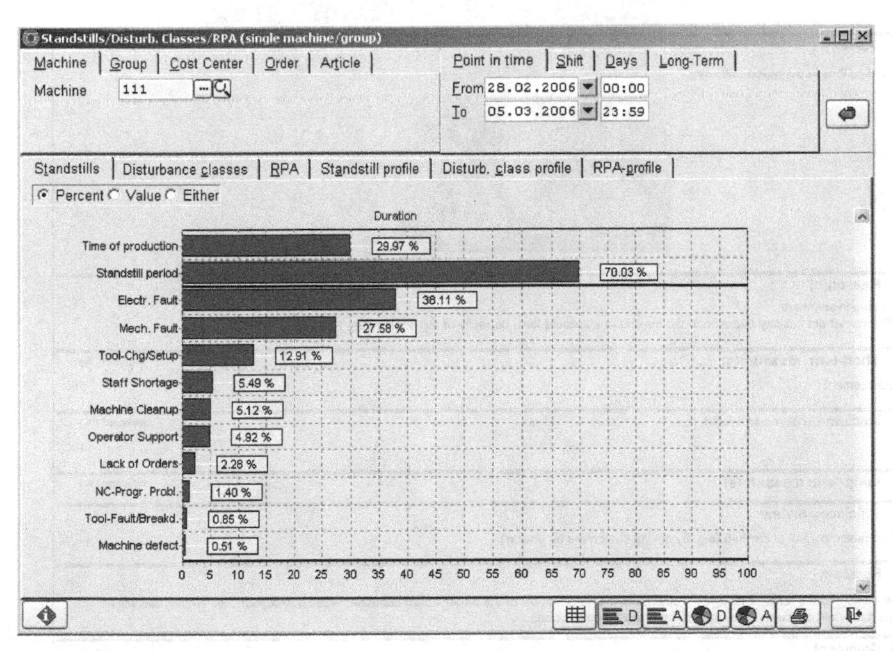

Fig. 2.3. Systematic errors displayed in the MES

2.2.2 Systematic failure processing

Every unplanned fault in the process must be treated like an internal complaint. All detected faults or errors must therefore be corrected permanently, which means that the organization will improve with each fault that has been detected and corrected. The long-term success of a company thus depends on a culture of improvement which has to be permanently anchored within the organization.

While this systematic approach of permanent and rigorous failure processing has long been practiced in companies today in the field of quality management, as regards the organization as a whole it is not yet state of the art, not by a long way.

Systematic failure management calls for interplay between different departments – it is a team activity: the foreman, the line engineer, the toolmaking, design or controlling departments make a contribution to processing the failure report. As regards the systematics of this, the 8D report, which was introduced by Ford, has now established itself as virtually the standard and yet can still be adapted or varied to suit the requirements of a particular company. What is important, however, is only that every detected problem must be corrected in a guided manner. Any statistics which

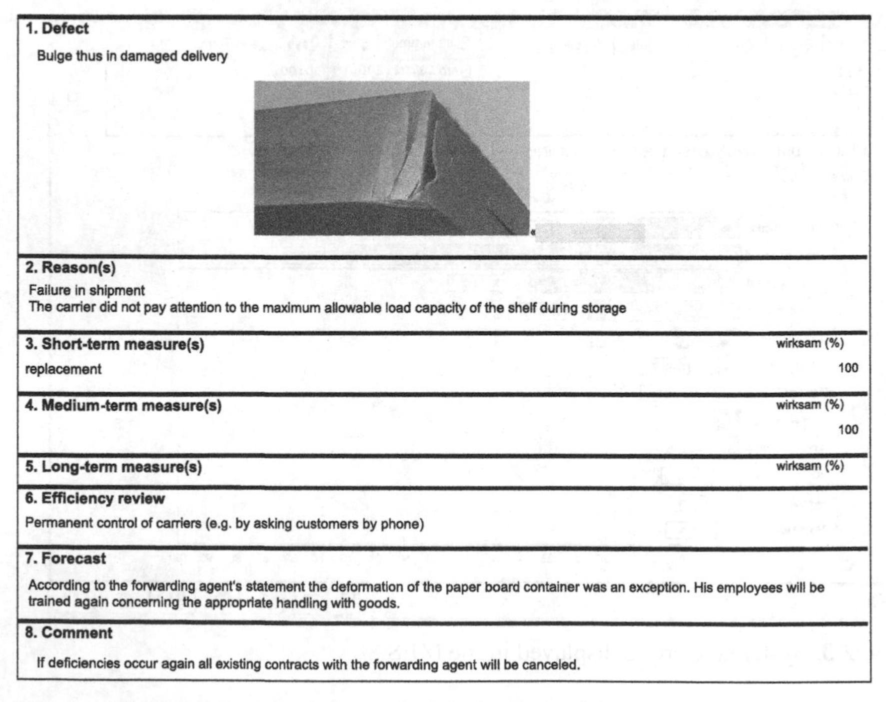

1. Defect

Bulge thus in damaged delivery

2. Reason(s)

Failure in shipment
The carrier did not pay attention to the maximum allowable load capacity of the shelf during storage

3. Short-term measure(s) wirksam (%)

replacement 100

4. Medium-term measure(s) wirksam (%)

100

5. Long-term measure(s) wirksam (%)

6. Efficiency review

Permanent control of carriers (e.g. by asking customers by phone)

7. Forecast

According to the forwarding agent's statement the deformation of the paper board container was an exception. His employees will be trained again concerning the appropriate handling with goods.

8. Comment

If deficiencies occur again all existing contracts with the forwarding agent will be canceled.

Fig. 2.4. The 8D report for systematic failure processing

do not automatically lead to failure processing are unnecessary and will thus disappear.

2.2.3 Tracking corrective action

Any failure, once it has been logged, can immediately generate a workflow in the MES system in which the processing steps (as, for example, in the 8D report) are specified and the departments processing the steps are defined. All actions to be carried out during the course of processing are then input into a mask for action tracking and can thus be monitored transparently. Not until all processing steps have been ticked off can the "case" be closed. Here the 8D report also requires a long-term prediction be made on the basis of the long-term corrective action taken (training, revision of drawings, checking of documents, and so on).

The workflow shown in Fig. 2.5 can be defined for different kinds of failure handling. In addition, an MES can also automatically generate

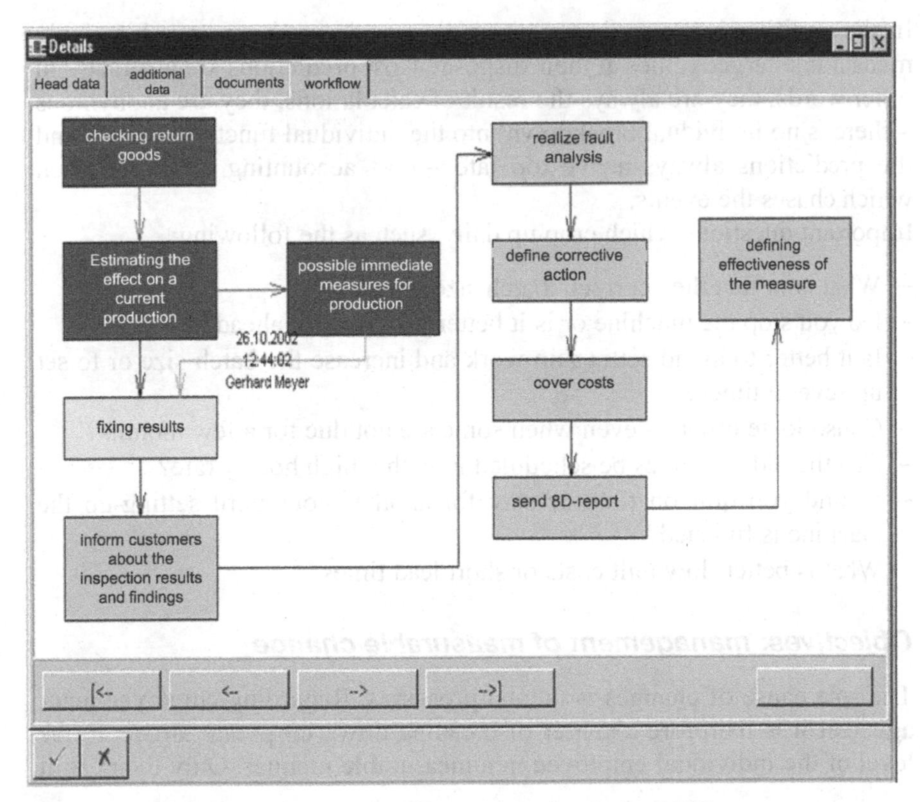

Fig. 2.5. Workflow for systematic failure processing in the MES

workflows for any kinds or instances of failure. To do so, intervention limits are first of all defined for particular operation sequences (machines, processes, work, faults, etc.) – for example, when a bottleneck machine is idle for more than 10 minutes. Then, should such a limit be exceeded, a workflow will be immediately generated automatically together with a serial number for the problem, the date and the actions required.

2.3 Process capability in collaboration

Collaboration may be regarded as process-capable when everyone in the company knows how he and the company can improve with respect to the customer. This will not be possible unless the real value drivers are converted into objectively measurable target values by which everyone in the company can work under monitoring.

2.3.1 Wasted work

In the traditional cost environment employees do not have any objectively measurable target values at their disposal. Cost predictions are abstract – in other words, they are always the result of calculations, they are anonymous – there is no individual breakdown into the individual function carriers and the predictions always arrive too late – cost accounting is an approach which chases the events.
Important questions which crop up daily, such as the following:

- What is in fact the “correct” batch size?
- Do you stop the machine or is it better just to work ahead?
- Is it better to avoid setting-up work and increase the batch size or to set up several times?
- Consolidate orders – even when some are not due for a few months?
- Can the old machines be scheduled in with a high hourly rate?
- Should you quit on time or stay for another hour until setting-up the machine is finished?
- What is better, low unit costs or short lead times?

Objectives: management of measurable change

The sole cause of changes is mental processes. Regarding employee management it is therefore a matter of breaking down corporate targets to the level of the individual employee in a measurable manner. Only then can it be ensured that the efforts involved in their collaboration are actually directed towards the process result.

Figure 2.6 shows by way of example how different corporate strategies (in other words, the answer to the question: with which customers do we want to stand out from the competition with which services?) can be formulated in internal targets for the production department and then in objectively measured quantities for the employees.

strategy/aims	key data	measurement categories	measures
cost leadership	economic efficiency	capacity	reducing downtimes and waiting times
quality	degree of purity	ppm value	operator inspection
on-time delivery	punctuality	deflation in days	reducing unplanned malfunctions etc.
flexibility	ability to deliver	cycle time in days	
market share	increase in Sales	increase in sales - amaount/ piece - in %	
design/ innovation	customers acceptance		(measures have to be worked out)
variants	range of parts	rejections	
delivery reliability	delivery accuracy	deviations	

Fig. 2.6. Breakdown of strategic targets on the value generation level

Only this gradual breakdown of corporate strategy down as far as the process level itself can ensure that the standards of the company with respect to its customers are living realities within the company as well. Not until there is a background of quantifiable target variables, such as reduction in process times, reduction in inventory levels, improvement in process reliability, reduction in scrap or set-up times, will it be possible to provide the employees with handy measurable targets which they can then use to improve themselves and the company. Such bottom-up decision-making structures call for a new kind of personnel management: in future they will be directed not by work instructions – as is the case in the traditional factory – but more and more by operational target agreements. In this way decisions will be more often made by those closest to the work itself.

The manufacturing scorecard as an operative component of an MES also offers the possibility of giving the employees (or the group) up-to-date information about "their" score (Kletti and Brauckmann 2004).

2.3.2 Operational target agreements

The targets and measured variables found and formulated for the value generation level are then grouped together into operational target agreements. These then form the logical foundations of a corresponding performance-related reward payment system.

Figure 2.7 shows by way of example how different target and measured quantities can be assigned to the unit or person responsible for the process. Group work is present anywhere several employees can jointly influence the process result.

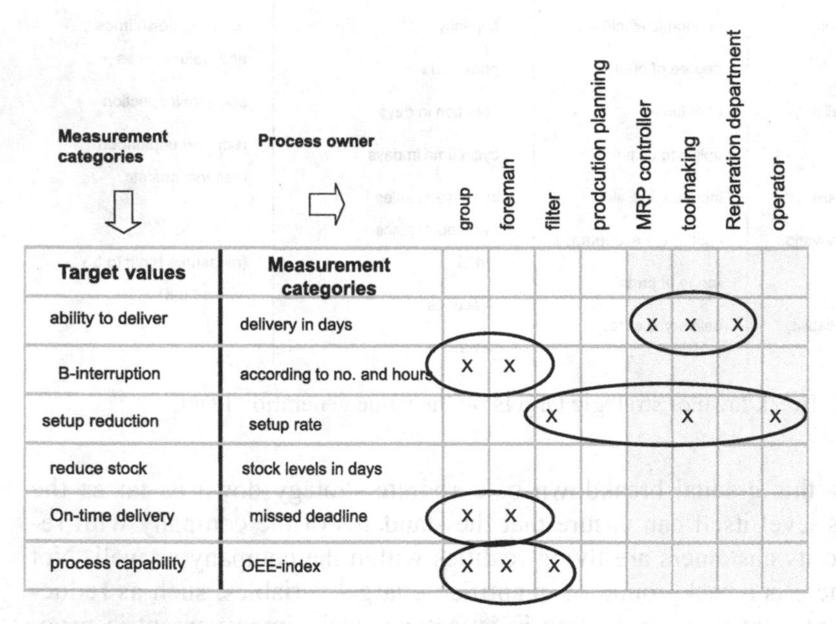

Target values	Measurement categories	group	foreman	filter	prodcution planning	MRP controller	toolmaking	Reparation department	operator
ability to deliver	delivery in days					X	X	X	
B-interruption	according to no. and hours	X	X						
setup reduction	setup rate			X			X		X
reduce stock	stock levels in days								
On-time delivery	missed deadline	X	X						
process capability	OEE-index	X	X	X					

Fig. 2.7. Assignment of measured variables to the unit or person responsible for the process

2.4 Process capability of information flows

Information and communication flows may be regarded as process-capable when they are in alignment with the value stream. The further the information flow separates itself from the material flow, the greater will be the slack-time and the ensuing losses.

2.4.1 The company as a paper factory

In the factory any piece of metal (machine, tool, mesh pallet, produced items, and so on) which does not have its accompanying data record is regarded as scrap in accordance with the certification rules. To give an idea of the sheer quantity of vouchers and the like required just for the flow of material: delivery note and works certification from the supplier (incoming goods); release decision, supplier appraisal by QS (quarantined store), stock receipt list (incoming goods store), job papers, material issue slips, time tickets, container marking slip, routing cards, batches, lot tracking, scrap report (production), externally processed order, container tracking (outlying departments), stores ledger card, label (storage facility), picking request, stock picking list (assembly/order picking), delivery note, invoice, shipping papers, QA documents (shipping), etc.

2.4.2 Interfaces without value generation

The reason for the paper factory is to be found in the fact that the information flows in the company are traditionally organized on a hierarchical basis – that is, in effect vertical to the flow of material. This can be seen in the organization chart which describes not only relationships of relative subordination or superiority but also instruction and reporting paths. Every box in the organization chart can be an individual or a function. The transition from one box to another is an interface and therefore in all cases a media discontinuity as well. Interfaces are cost-intensive, highly time-consuming and a quality risk with high error probability.

Each interface produces paper. Companies pay enormous amounts of money for their paper production. The printing costs alone of larger companies are today already estimated at around 5% of sales volume. Even the "normal" passage of an order through production calls for a high level of written output, starting with preparation of the works order and printing hard copies of the job papers such as time tickets, material issue slips, trial orders, scrap reports, schedule cards or routing cards.

Each document has an expensive career: it is created somewhere (usually where it is not needed), sent somewhere (to a foreman, for example), from here it is distributed (to the machine operator), filled out by him, returned, either glanced at or checked over and signed by the foreman, collected together and sent back to the office/EDP department, input at one time or another (EDP or MS Excel), and finally, once everyone has forgotten all about its contents, evaluated. Copies are made of the evaluation, these copies are sent out for distribution and finally everything is filed away. This is a long chain of activities which ties up time and personnel.

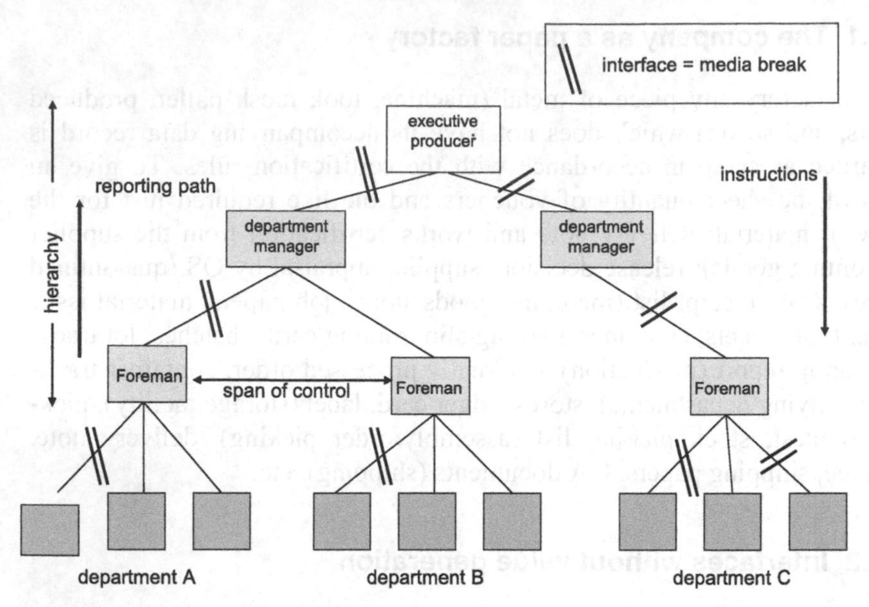

Fig. 2.8. The organization chart as a traditional communication model

In addition, each document calls for different knowledge and thus means different responsibilities. Responsibilities in turn require complex coordination activities between the units involved: informing, checking back, assigning, passing on, etc. Communication between different departments always means that the corresponding paperwork has to be provided as well. Not only that but in nearly every company the most varied regular reports are issued, reports which are expensive to prepare, whose original purpose has become lost in the mists of time and which are therefore never read but immediately filed away untouched.

Another and a less often considered point of weakness in decentralized information structures is found in the transformation of the existing knowledge which is basically present in the company but not currently available: it is found in drawers, folders, files or in the heads of the employees.

2.4.3 The way to paperless production

The alternative to media discontinuities is a dialog-based decentralized access to central data storage: the right information at the right time in the right place. With this, all data hoards such as lists, statistics, files, various written records and so on can be broken up at a stroke and the really necessary information made available to all users in real time and without the use of paper.

Grouping together and networking all of the relevant business processes within a data storage facility automatically eliminates the interfacing costs found in the classic organization which arise from the isolated concurrent and consecutive arrangements of business processes which are bound together in their results. An essential requirement of this new approach is that all of the processes necessary to value generation are networked in a data model. Modern production data management automatically captures the value-adding processes in the background and makes the data available – in a manner of speaking as a waste product – decentrally to all information end-users from a central production database. This shared use of central data which are only held in a single location represents a revolution when compared with the usual organizational forms which we have just briefly sketched.

Automatic online data acquisition is therefore a requirement of decisive importance to the integration of business processes which within the organization based on division of labor cannot be implemented economically or efficiently by means of written records. For example, recording job processing on a centralized basis means that information can be obtained automatically and simultaneously: order progress data for production control; machine problems for time scheduling, beginning of setting-up for the next processing step, supply dates for the supplier, delivery lead time for the customers, processing times for the controlling department.

The information hidden behind every business process is available in real time via a central data storage facility with decentralized access: the

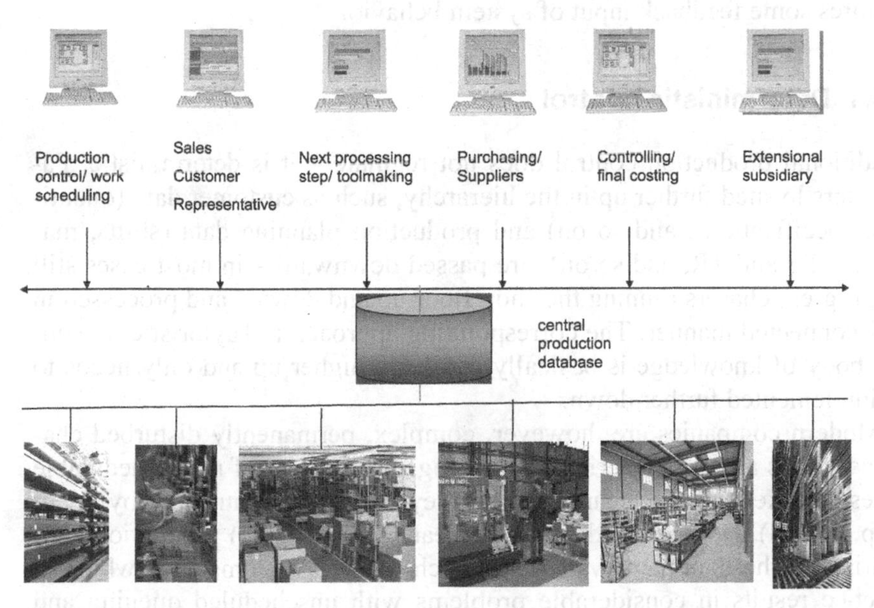

Fig. 2.9. Process-capable communication model

right information at the right time in the right place. In the new factory, machine states, order progress, time scheduling, batch tracking, preventive maintenance, tool management and statistical cost accounting are viewed and presented as networked events. Instead of personnel organizations based on the division of labor together with their traditional departments and hierarchies, processes are made available in real time by networking all information consumers. The right information at the right time in the right place. This means that even traditional management structures such as hierarchies and departments can be replaced by information structures which all employees are networked and have access to the same information.

The concept of information networking has some other effects of decisive importance:

- The supervisor as a source of information for the employee now becomes less important.
- The company's knowledge base which is normally somewhere or other in an unmonitored form (files, drawers, individuals) suddenly becomes available.

2.5 The process capability of flow control

Flow control can be regarded as process-capable when it is based on closed-loop control. In contrast to open-loop control, closed-loop control requires some feedback input of system behavior.

2.5.1 Deterministic control

Traditional production control does not regulate – it is deterministic. Parameters located further up in the hierarchy, such as customer data (quantities, specifications, and so on) and production planning data (shifts, machines, TE and TR, and so on), are passed downward – in most cases still by progress chasers running the shop floor up and down – and processed in a disconnected manner. The corresponding approach is Tayloristic: the entire body of knowledge is basically available higher up and only needs to be implemented further down.

Modern companies are, however, complex, permanently disturbed chaotic systems also characterized by having a high share of non-schedulable times. Unlike simple (linear) systems they cannot be controlled (by open-loop control). Notwithstanding this, linear (deterministic) production control is still the dominant system approach today, a circumstance which in practice results in considerable problems with unscheduled queuing and waiting times, unplanned inventory, deadline infringements and the resulting extra costs in production (a hectic atmosphere, overtime, special shifts, express transportation, and so on).

2.5.2 Control with feedback (closed-loop control)

Only by feeding back actual system behavior it is possible to have closed-loop control. Incorporating a feedback loop in the system is thus the decisive precondition for the process capability of the material flow. This in turn means that the elements of the control loop (set-point values, control variables and measuring elements) must be embedded in the organization.

An efficient flow control system must be able to react to faults flexibly and with minimal time lag. To ensure that there is a fast and competent reaction to faults and malfunctions the control loops need to be organized appropriately.

This arrangement results in an organization model in which the planning level (production control) on the one hand and the material planning level (actuator) on the other are linked together via a measuring element (MES/decentralized online planning table). Unlike the traditional, deterministic Tayloristic flow model with central production control and pro-

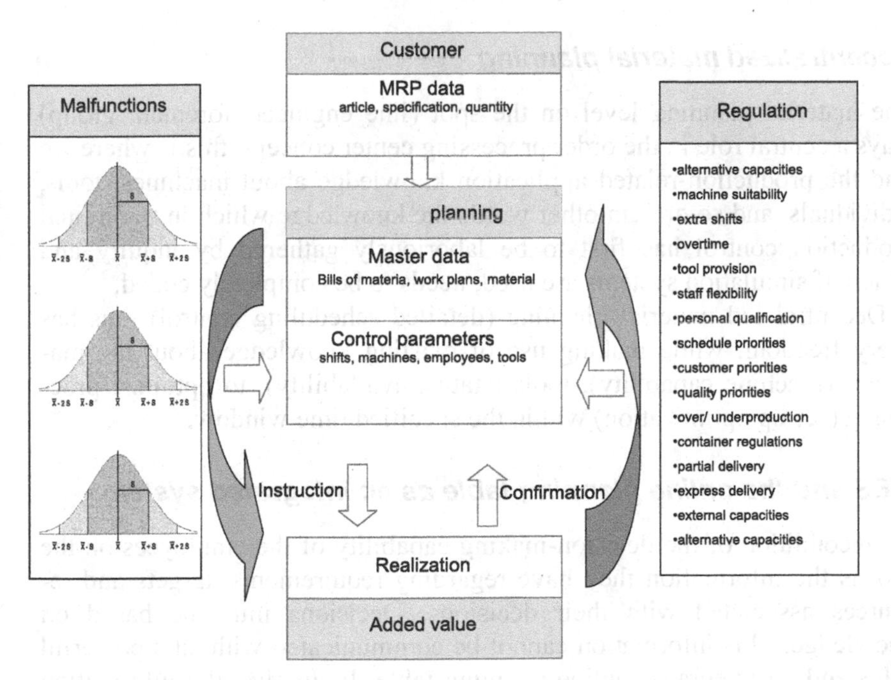

Fig. 2.10. Production control as a control loop

gress chasers, a modern order processing center is characterized by a process-oriented distribution of tasks.

Central planning

The order processing center is the central process owner. In a similar way to the head of department who bears a responsibility for the department, the process owner has the responsibility for the process results. In contrast to the traditional vertical style of department classification, here the process owner is biased horizontally by the process orientation – he thus forms a part of the process chain (supply chain).

The order processing center as process owner thus has a responsibility which ranges all the way from the external supplier via outlying units and as far as the customer. It is furnished with the corresponding powers: it can agree on dates with the supplier without involving the purchasing department and is also the responsible contact for the customer with regard to delivery dates without having to involve the Sales department.

On the basis of the information about bills of material, operation sheets, order quantities and shelf inventories, the ERP system, as part of flow scheduling, determines the basic dates (earliest and latest starting dates and also the earliest and latest finishing dates) for each operation and these are then passed on to the material planning level as requirements.

Decentralized material planning

The material planning level on the spot (line engineer, foreman, group) plays a central role in the order processing center concept: this is where we find the production-related application knowledge about machines, tools, individuals, and so on – in other words the knowledge which in traditional production control has first to be laboriously gathered by inquiry and which, if simulation systems are used, needs to be completely coded.

Decentralized material planning (detailed scheduling control) thus has every freedom, while making use of existing knowledge about the machines (machine capability), tools (status, availability), to optimize processing (set-up optimization) within the specified time window.

MES and the online planning table as an integrated system

A precondition of the decision-making capability of the employees on the spot is the information they have regarding requirements, targets and resources associated with their decisions. Decisions must be based on knowledge. This information cannot be communicated without a powerful MES and an integrated online planning table. It supplies the information for decisions about finite scheduling, control of process flow, material transportation, trouble-shooting, quality control, maintenance, and so on.

In traditional production control the planning level – today usually located within the PPS/ERP systems be gained – and the execution level function alongside each other without being monitored. In this the progress chasers, time keepers or foremen form a parallel organization which reveals itself in schedule meetings and various adjustments which take place outside the data model and then have to be input back into the system afterwards. This is an approach which in practice is called peer-to-peer communication.

The modern MES now makes it possible to integrate planning and execution in a single system without interfaces. The information in question is passed on online:

- Work instructions go out via the network to the terminal close to the machine. In this way all further associated information, whether from the ERP database, the QS database, the DNC server, and so on, is automatically made available electronically (and can still be printed out on the spot if necessary).
- Status reports are also provided online regarding the work centers/machines, this taking place in step with the work cycle. This means there is no need for written records with their usual risks such as uncertainty, arbitrariness, time lag, incompleteness.

Figure 2.11 shows by way of example a graphic production control station as an integral component of an MES. It offers the possibility of, on the one hand, taking over planning inputs online from the ERP system from "above" (such as articles, operations, set-up and processing times, use of material, production quantity and customer deadlines) and on the other hand comparing them online with the current actual situation from "below" (such as machine states, dynamically calculated lead-time remainders, order progress, loading horizon, capacity conflicts) – in other words, from the operative level. With this closed control loop the planner has for the first time the ability to react without slack-time – in other words, to regulate.

Special importance should be attached to the real-time updating of the database (master data in the ERP system): as has already been stated, the reliability of the database is the decisive requirement for the efficiency of material flow. Manual care and maintenance of the input requirements (operation sheets, processing times, set-up times, machines, and material requirements) is in practice not possible, especially against a fast changes actual situation.

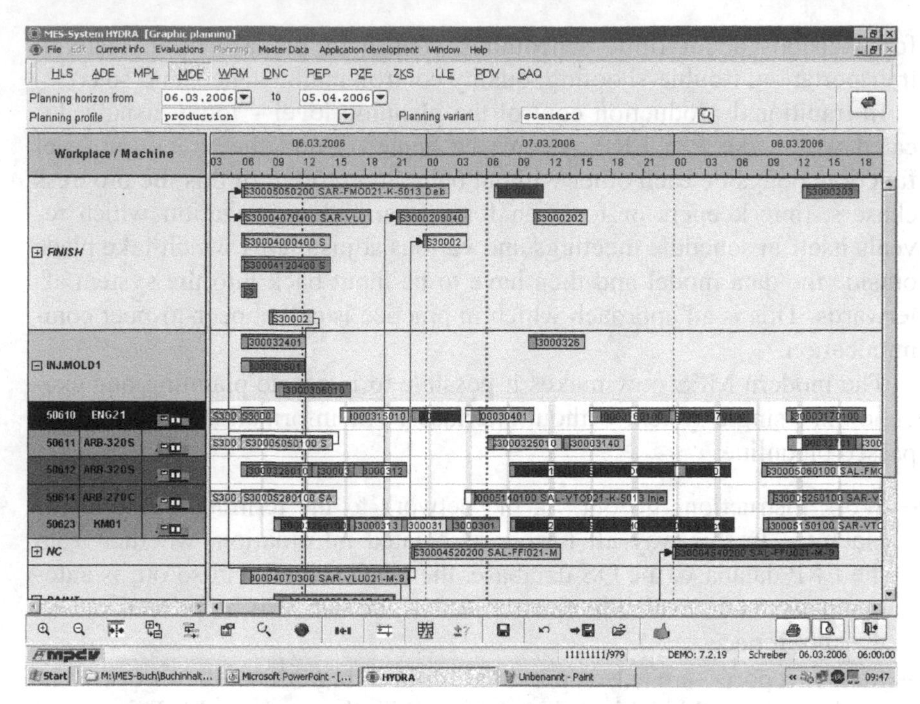

Fig. 2.11. Online control station as a tool for material planning

2.6 Summary

The nature of competition is shifting increasingly from competition between products to competition between business processes. This makes the demand for process capability the very highest priority for the new factory. Process capability means the company and its resource deployments are oriented rigorously to the process results and thus to the customer.

Traditional production attempted to guide resource consumption in processing by means of the counting mechanism of the company accounting department with unit costs used as the target quantity. Unit costs in the sense of costs assigned to the corresponding originator of the cost hardly exist any more today. The modern service economy is characterized by a high and still increasing proportion of overheads. Overheads cannot however be assigned to individual cost units by any logical procedure. The result of this is that in the past there was no possibility of evaluating added value on the market (and thus for the customer) with the costs incurred allocated source-specifically.

The ensuing lack of an effective control loop between the market and the company (price-based control) in practice results in enormous waste. This manifests itself in, for example, the fact that in virtually all companies the possibilities of saving on costs have been exhausted although these instances of waste do at the same time represent a considerable process potential.

The process-oriented approach laid down in the certification rules with its demands for customer orientation, process control and a culture of permanent improvement makes a systematic identification of processes necessary. It is, in fact, part of the central weakness of traditional cost accounting that it cannot carry out this identification on a methodical basis: it lacks a time dimension.

Process identification means concurrent in-process data acquisition for all value-adding processes in the company and this task can only be solved economically by having a concurrent MES system running in the background. For this reason, coherent process identification is not feasible without modern information technology in the form of an MES system.

Although in the meantime an overwhelming proportion of all companies has been certificated in accordance with the process-oriented standards ISO 9001 or TS 16949, this approach is not actually being put into practice. The resulting process capability is not state of the art as it is understood today.

Table 2.1 When is a company process-capable?

Process capability of the organization	An organization is process-capable when it is consistently self-teaching down to and including the operative level. As in aviation, where every flying accident makes flying safer, every error must make the company better.
Process capability in collaboration	When every employee on the operative level, thanks to access to quantifiable requirements and targets, knows how good he is and how he can improve himself and the company.
Process capability of resource deployments	Resource steering is process-capable when it is oriented by process parameters (lead times, target dates, inventory levels, etc.) instead of by false key data such as unit costs, batch sizes and hourly rates.
Process capability of IT processes	IT structures are process-capable when they reproduce the value stream without the use of interfaces – in other words, without media discontinuities in the form of papers, written records, vouchers, planning extracts and so on.
Process capability of controlling/reporting	Reporting is process-capable when statistics have been replaced globally by reviewing.
Process capability of flow control	Flow control is process-capable when traditional centralized open-loop control is replaced by decentralized closed-loop control.

Literature

ISO/TS 16949 Technical Specifications, 2nd edn., 2002

Hendersen B (1974) Die Erfahrungskurve in der Unternehmensstrategie [The experience curve in corporate strategy]. Herder & Herder, Frankfurt/Main

Kaplan R, Norton D (1997) Balanced Scorecard. Schäffer-Poeschel, Stuttgart

Kletti J, Brauckmann O (2004) Manufacturing Scorecard. Gabler, Wiesbaden

Rehbehn R, Zafer B (2003) Mit Six Sigma zu Business Excellence [With Six Sigma to business excellence]. Siemens AG, Berlin and Munich

3 Added value from software

3.1 The company as information system

3.1.1 Information as a production factor

Modern companies are predominantly information-processing systems. It can be assumed today that more than half of value-added costs flow into the production factor of information. Production itself is losing strategic importance more and more rapidly. This manifests itself in, for example, the fact that companies are relocating their production abroad for reasons of cost or are reducing vertical integration without losing their competitiveness or even actually improving it. This is also confirmed by a survey conducted by the VDMA which revealed that a vertical integration of almost 50% in 1998 had shrunk to nearly 40% in 2004, coupled with a simultaneous improvement in position in international competition.

Production is being increasingly replaced by the service capability of offering the market a wide range of product variants to suit customers' wishes while at the same time ensuring a high quality of products and services as well as an excellent delivery service. The features listed here, such as conformity with customer wishes, services, quality, range of product variants, delivery service, are none of them properties which can be secured via the traditional concept of production and thus to be pinned to the product in a measurable manner. They are primarily based on information processing and the ability to have the required information available at the "right time", in the "right quantity" and at the "right place". Command of information management along the value chain is becoming more and more important for the competitiveness of companies whether they now manufacture physical products, such as the capital goods industry, or virtual products, such as for example the software industry.

The more the value creation of a company for the customers consists of combinations of products and services (with services taking an ever greater share), the more productivity a company will have to invest in its information processing and thus in the deployment of supportive software. This does not mean however that investment in software will automatically add benefits in productivity, flexibility or transparency. A large number of examples spring to mind in which a half-hearted and unsystematic use of software tended to bring disadvantages with it instead. Software alone does not bring added value – not until software is installed in the right environment and used systematically will added values be achievable.

3.1.2 Re-engineering and integration

When you examine the value-adding processes in a company, what they all have in common is that they are accompanied by information which on the one hand documents the status of value generation and on the other hand describes the performances which still have to be carried out. Information is thus the real process driver and in this way controls operational sequences in the company. One obstacle to fluid processes is, however, forms of organization which are oriented by performance which hinder and slow down the processes on the basis of departmental boundaries. A further obstacle to fluid processes are the innumerable media discontinuities which cut up the information into large numbers of individual parts. This makes it very difficult to track, control or direct these processes.

There is no doubt that overcoming a conventional, Tayloristically biased organization structure is a management task which goes hand in hand with considerable changes in the culture of the company. If information processing makes up the greater part of value generation, rationalization will have to be applied to the information processes. That is the idea behind re-engineering. Re-engineering aims at a restructuring of the information processes with the objective of obtaining control over increasingly greater demands with regard to quality, service, flexibility, costs, target dates and delivery times. The decisive changes will take place in the future not in the field of technology but rather in the definition of and control over information processes. The prevention of media discontinuities is, on the other hand, a task which can be carried out with the aid of technical tools. Here the control and processing logic is mapped into the software. When internal and external data networks are used, the information can be passed on in real time over departmental and corporate boundaries. It is of decisive importance here that the systems involved actually understand each other. The situation should not be permitted to arise whereby the information is being exchanged digitally but nevertheless, due to a lack of or incompatibility in data interfaces, a human agent must actively step in as a mediator of information.

These two aspects – modernization of the company organization by re-engineering and optimization and software support for information flows – must go hand in hand if a sustainable improvement in the value processes is to be achieved. It cannot be pointed out too often in this regard that management is required here to involve itself actively in the successful realization of the corresponding systems. Many projects come unstuck because management only feels itself responsible for allocating means and resources and the persons in charge of the project fail not in the technical implementation but rather in the necessary organizational and personnel-related design work.

3.1.3 Information processing in production

If a comparison is made of progress in information processing in the various functional areas of a company, it will become clear that production in particular is still frequently suffering from deficiencies in information processing and networking. Production steps are characterized by an increasing complexity which is caused or at least influenced by high product variance and customer-specific implementations of the products. This impacts especially on the capital goods industry whose particular challenge is the economically efficient production of a batch size of one. Mastering complexity while simultaneously securing productivity represents precisely the ideal typical conditions for the use of modern information processing.

It is astonishing in the light of this that a paper-based exchange of information is still prevalent in the production departments of a large number of companies. Paper-based collection of data for machine run times, machine availabilities and OK parts from actual production in progress is a living anachronism which is still a routine sight in industry. These activities are amongst the most cost-intensive, non-value-adding activities in the production environment which industrial companies still indulge in today. This way of working is, however, not only extremely inefficient but it actually encourages errors and inaccuracies. Furthermore, it should be noted that the employees are also measured and judged by the data and information arising in the production process. Payments systems based on the evaluation of the quantities of items produced will automatically bring with them the danger of people trying to manipulate them.

It is not only operations planning which is brought into confusion by the errors which can occur in manual data collection. Even today machine-hour rates are used as the basis for calculating selling prices. If, in spite of all its known deficiencies, this method which takes the machine-hour rate as the basis of costing is still used, at least every effort should be made to prevent the corresponding information from being falsified by human error or weakness.

3.1.4 Machines as information-processing systems

The theme of information processing primarily brings to mind software running on classic IT hardware such as mainframes, servers or PCs. However, in parallel with this world of company IT, a world also exists of automation technology and machine-oriented software. Machines and installations or the automation technology used in them are themselves in most cases complex information-processing systems. Technical functions in machines and installations which were once implemented by means of mechanical and specialized electrotechnical components are today more and more often based on software and standard IT. In this way software, via digital sensors and actuators, regulates and controls the movements and operations of the machine while industrial-grade PCs serve the machine operator as a communications interface with his machine, with higher-level software systems or, via the internet, with the outside world. With modern information-controlled components or machines, the part of unit costs made up by the software can easily be as much as 25% to 40% or sometimes even more. Nor is it a rarity for software development spending to amount to 30% or more of total development costs – something which is also reflected in the constantly increasing number of software developers employed in companies.

The primary reason for the growing proportion of software not only in capital goods but also in consumer products such as cars, entertainment electronics, telecommunications is to be found in the fact that software allows the products to be tailored much more simply and flexibly to the specific requirements of the customer and is also able to offer additional new services and features. Even with traditional products, the use of software is therefore more and more becoming a competitive factor, the basis for customer-oriented added value.

From the view point of the MES a considerable problem existed in the past in that information processing in automation had developed independently and interfacing with machines was both complex and product-specific. The special requirements applicable in the automation environment, such as real-time processing, safety, availability and even costs, have resulted in special, mutually incompatible controllers, bus systems, operator terminals, data storage facilities and programming languages. As standard IT and software is used more and more often, even in automation, interfacing problems between the two worlds of the company are declining and this is making it possible to implement standardized and considerably more efficient information and communication processes.

3.2 MES in the capital goods industry

Using software to increase performance in information processing is in itself nothing new for the company and has been an ongoing topic for years. What is new, however, is that the question of software application arises again and again in response to new technical aspects and new possibilities of using it. Due to the high speed of innovation in the IT industry the basic technologies are in a constant state of change and new areas of application are constantly being opened up. (Marketing statements such as "This software will secure your future" or "Solve your problems with this software" are intended to win potential customers for new software investments.) However, after years of euphoria we can now see wait-and-see attitudes, particularly in mid-sized companies. It is precisely with new technologies (such as MES, for example) that a certain skepticism is entirely advisable – as also an analysis of the environment – before any actual investment in software is made – but then it should be implemented systematically and with a clear direction defined.

3.2.1 Characteristics of the capital goods industry

The capital goods industry, which in our view includes in particular mechanical engineering, plant engineering and construction and electrical engineering, is currently once again one of the cutting-edge branches of industry. Following the lamentable failure of the "new economy", people are once again remembering traditional areas of strength. Strong growth in sales, high export shares and technological leadership are some very presentable characteristics of this sector of industry. Despite what tend to be unfavorable influences in the environment, such as unfavorable exchange rates or higher costs for energy and raw materials, the industry's competence in problem-solving is winning through in the international arena and demand for the products is at record levels worldwide.

Notwithstanding this, this branch of industry is not resting on its laurels at all. As the VDMA's 2004 survey of trends indicates, as a way of improving the competitive situation, action in the fields of "forced product innovation" and "employee qualification" is currently very popular. To gain competitiveness in a location with high wage levels, it is necessary on the one hand to defend product leadership and on the other to use human resources more and more efficiently. A decisive lever in accessing these benefits is IT solutions. This is all the more important in the light of the fact that a shortfall in qualified personnel for future years can already be noted for this sector of industry. It is therefore a matter of automating routine tasks more extensively and giving complex operations support from

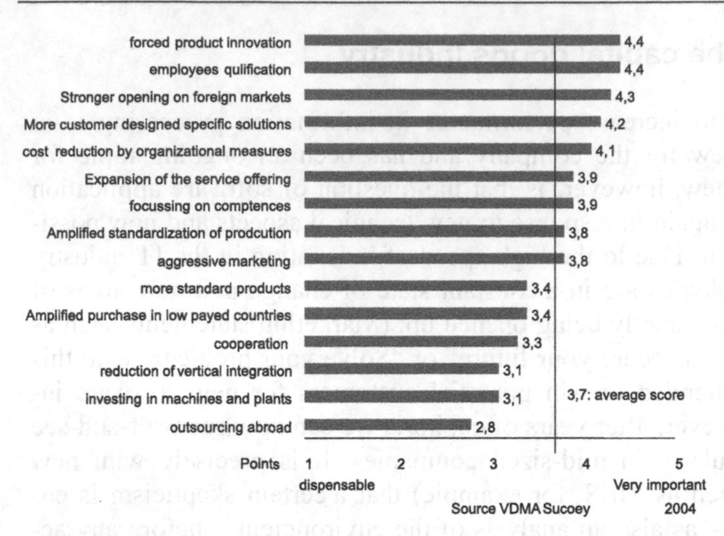

Fig. 3.1. Innovation survey

suitable software instruments. That companies in our industry really do want to take this road can again be seen from the above-mentioned trends. Here it becomes evident that as a strategy for improving the competitive situation the option of relocating production abroad is regarded as well below average and takes last place amongst all the possibilities!

3.2.2 MES in the IT software landscape

In companies in the capital goods industry, MES is as a rule surrounded by a large number of supplementary software solutions. How these modules are classified will depend on the one hand on their functionality within the context of value processes. Externally these are the company's suppliers and customers, internally the functional areas with their specific tasks within the process. The processes can be roughly subdivided into commercial and technical tasks. As regards the integration of the MES system into this software environment, different requirements will apply. All in all it can be said that the closer you get to the production process, the greater is the necessity to integrate the software products with the MES. Here clear requirements relating to interfaces arise particularly for the classic ERP components of inventory control (DIS) and production (PPS). In addition, the tight connection with the level of automation is to be emphasized. Here the old method whereby the ERP has a unidirectional supply of machine data from MDC (machine data collection) or from PDA (production data

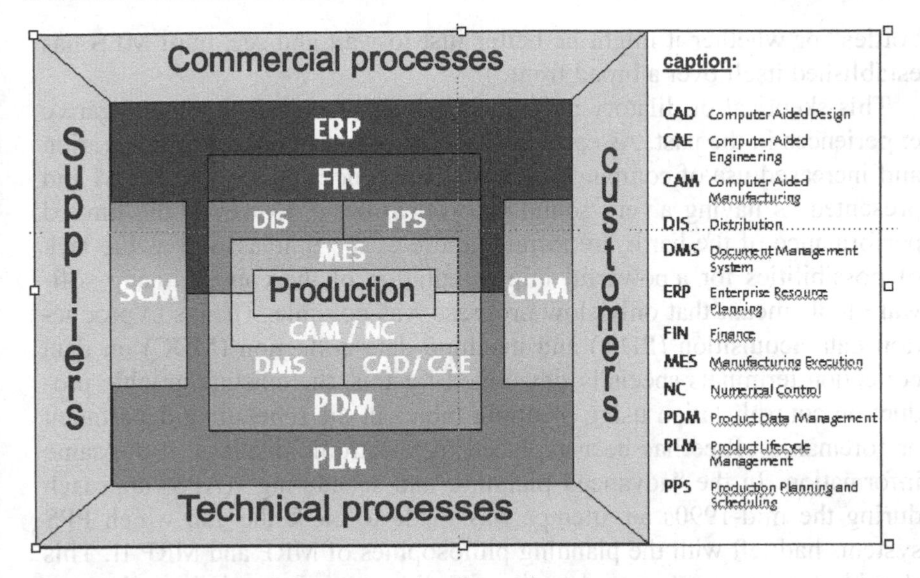

Fig. 3.2. Integration of the software landscape into the company

acquisition) has now become a dialog between the MES on one side and these systems on the other.

The main aim of the close connection to inventory control is to provide support for the detailed scheduling control of the production materials and components which are available for the production orders. This brings one of the most indispensable requirements for production into the planning process, namely material availability.

In the past PPS was often incorrectly rendered in German as "Produktionsplanung und Steuerung" (in other words, production planning and control). In fact, not giving the production scheduler a possibility of control is the precise inability which is characteristic of PPS systems. The classic approach here is to plan against unlimited capacities and possibly even to use starting dates which are already in the past. The MES can deploy its strong points here of backlog-free planning on the basis of existing capacities and the current progress of production being reported back in real time. On this basis PPS can then release customer orders on hand to production, doing so on a rolling basis; the systems run in synchrony and supplement each other.

3.2.3 MES in the technology life cycle

Despite the very respectable benefits of MES this concept is still far from winning through in the investment goods industry. In some cases there is still a great deal of skepticism as to whether MES is just "old wine in new bottles" or whether it might be better just to wait and see, until MES has established itself over a broad front.

This skeptical or dilatory attitude is primarily rooted in some negative experiences in the past. As early as 20 years ago the concept of integration and increased use of computers was taken up under the keyword CIM and presented as having a very sound theoretical basis. However, the limited performance of the hardware infrastructure of the time as well as the lack of possibilities for a powerful implementation of the concept using software tools meant that only slow progress was possible. Thanks to production data acquisition (PDA) and machine data collection (MDC) on data collection terminals specially developed for this, the existing graphic production control station using planning tables in the scheduling department or foreman's office has been replaced by electronic displays of the same information. In the "advanced planning and scheduling" (APS) approach during the mid-1990s an attempt was made to close the gap which PPS systems had left with the planning philosophies of MRP and MRP II. This should now have put an end to the situation whereby periods in the past were accorded planning relevance with regard to production steps not as yet carried out. Nevertheless even the APS method could not get beyond being simply a more refined planning attempt to close the control loop between ACTUAL and TARGET.

With MES the integration of all information relevant to production – in other words, to the areas of personnel, material resources and production resources – has now been completed. Data about available personnel resources with qualification profiles, available raw materials and components, and also the free machine capacities including the tools and equipment required, all flow into the MES. At the same time, manufacturing progress is reported back continuously to the planning system thereby closing the control loop of production control in such a way as to make iterative replanning in production possible on this basis.

In retrospect, MES, like its predecessors, has run through a development curve typical of new technologies which corresponds to Gartner's hype cycle. If the idea of the CIM is taken as the starting point, with MES we have now reached the "slope of enlightenment". A growing number of systems are available on the market which are technically mature enough for practical service and whose utility value has been demonstrated by concrete implementations.

It does therefore appear very likely that in the foreseeable future MES will play a considerable role within the context of production organization and control and will also reach the plateau of the application level and widespread industrial application. Ultimately potential users will have to decide whether MES will prevail or succumb to its own hype.

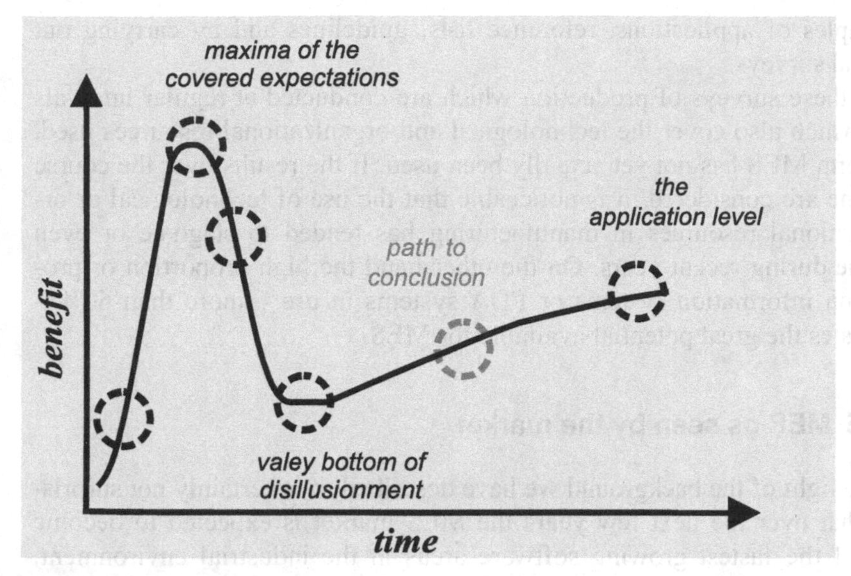

Fig. 3.3. Gartner's hype cycle

3.2.4 MES as seen by the user

A frequent indication that a technology hype bubble is bursting is when its technical terms are ridiculed or trivialized. In Germany, CIM, for example, gets called "Simsalabim" ("open sesame") and BtoB "to be or not to be". With new IT terminology there is always the danger of them being picked up by a large number of vendors in an attempt to exploit the market potential and who then redefine the terms for their own purposes. The result of this is a considerable lack of transparency for potential users. It is for this reason important that organizations such as MESA, NAMUR or VDI endeavor to bring clarity to the definition of the term MES and to define the functional framework of MES.

In Germany MES is still a comparatively recent term. In a survey of 670 companies in the manufacturing industry conducted by Trovarit in 2004 the term MES was unknown to more than 50% of the companies while only 7% indicated they were familiar with the topic. Comparable situations may also be observed in other fields of software technology as well, such as PLM or the digital factory. In mid-sized companies in particular there is in most cases great uncertainty as regards areas of application, practical implementation or even the economic benefits of MES.

The VDMA, as representative of a broad range of small- and medium-sized businesses, has therefore committed itself, particularly in the case of new technologies, to clarifying matters for users, doing so by presenting examples of applications, reference lists, guidelines and by carrying out special surveys.

In these surveys of production which are conducted at regular intervals and which also cover the technological and organizational resources used, the term MES has not yet actually been used. If the results over the course of time are considered, it is noticeable that the use of technological or organizational resources in manufacturing has tended to stagnate or even recede during recent years. On the other hand the high proportion of production information systems or PDA systems in use – more than 60% – indicates the great potential available for MES.

3.2.5 MES as seen by the market

In the light of the background we have described, it is certainly not surprising that over the next few years the MES market is expected to become one of the fastest growing software areas in the industrial environment. The corporate consultants of ARC forecast, for example, an annual growth of 11% for MES in the process industry.

What is difficult with prognoses of this kind is that there is no clear structure to either the requirements made of the systems nor the vendor environment. The MES supplier market is still characterized by a high degree of heterogeneity and is addressed by several groups of suppliers, something which is explained by MES's positioning as a link connecting ERP and automation.

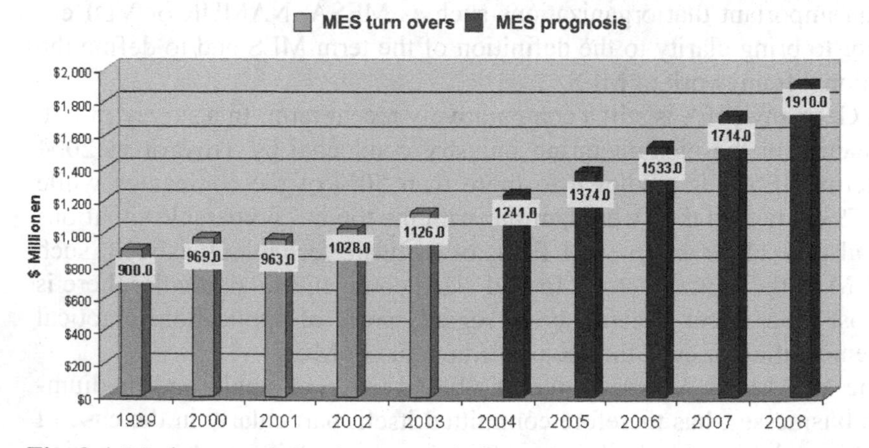

Fig. 3.4. Market growth for the process industry (source: ARC)

Due to the increasing demand for MES many suppliers of automation systems have expanded their range of products and services in this direction. There is no doubt that the marked commitment of the automation suppliers will result in further expansion and improvement of interfacing and communications between automation systems and MES systems. The interest that ERP suppliers have in MES is to further close the information gap which has so far existed in manufacturing, doing so either by developing their own solutions or by cooperating with standard MES vendors. A functioning connection with the higher-level ERP system is an important and critical factor in an MES system and defining practicable and standardized integration interfaces for ERP is an important task in the further development of MES. The engagement of the ERP suppliers who have already widely established themselves in the industry could be a decisive influence on the development of the MES market.

The third group are the actual MES specialists who can already offer mature, self-contained solutions for the MES market. Where differences exist they are mostly found in the scope of MES functions covered and in the primary orientation towards specific sectors of industry or types of company. Due to their years of experience and their customer-oriented flexibility these suppliers will continue to play a decisive role. Companies who are looking into MES and planning to introduce an MES system should abandon the idea that they can buy an MES system "off the shelf". MES must be tailored to their own particular situation and may very well be spread over a number of different software systems. Companies should take an eagle's view on their own manufacturing layout and create a structured master plan for their own MES system selection.

3.3 Preparations for MES-implementation

Continuous changes cannot be possible without continuous thinking processes. Systematically breaking down the performance promise of the company as expressed in its targets into key data and measured variables and going right down to the operational level will trigger a long-term dynamic process which cannot be laid down in advance nor even cast in an action catalog. For this reason it makes sense to work out the structures, objectives and room for maneuvering, doing so in collaboration with the employees. Only with this collaborative procedure will it be possible for the company to identify itself via the process responsibility of every single one of its employees.

3.3.1 Identification of objectives

A systematic application of MES will result in major consequences within the company. Benefits of MES nevertheless are not always understood in advance by those affected:

- The organization becomes self-learning, something an MES makes possible by its automated workflows and automatic escalation functions.
- The personnel field becomes decentralized and many decisions are made on the spot by a bottom-up process and then communicated upwardly.
- Leadership on the shop floor is carried out via KPI-based performance data by which the employees orient themselves at all times.
- Employee remuneration models are now based on processes and no longer on static figures.
- Central data storage coupled with decentralized data availability is revolutionizing the classic factory IT infrastructure by eliminating media discontinuities – value generation becomes interface-free.
- The concept of hierarchy is changed by process-capable information structures: the hierarchy becomes a service provider.
- Due to the introduction of MES, flow control, traditionally occupying a central location, turns into decentralized material planning with close-to-process control loops.

Within the context of an MES deployment project it is therefore important to draft the most realistic model possible for a future manufacturing organization. To do so, the existing situation will need to be carefully analyzed, the potentials for improvement identified, and on this basis the measures required for implementation obtained. The best course cannot be laid without clarity about its destination:

- What strategic direction should be implemented?
- What objectives support this corporate strategy?
- What key data and measured variables can be developed for these objectives?

3.3.2 Systematic process development

The permanent orientation of resource deployments by the process result calls for systematic process development involving all process owners. While decisions relating to economic efficiency were made in the past solely on the basis of figures from the accounting department, tomorrow's factory will, with the aid of an MES system, relate its key performance figures to the operational sequences behind these figures – the processes.

This far-reaching change affects the entire organization and it must therefore be borne by all employees.

Unlike the traditional procedure used with changes, which were implemented by the top-down method, here the process owners who are directly affected should be addressed. Again and again, on-the-spot workshops have demonstrated that the workers as a rule are the ones who point out recurring deficiencies in the organization. In the past they never had a system to express what these negative situations mean in terms of time and money, they never could quantify them or even eliminate them. Here the lost potential for bottom-up improvement is comparatively enormous: failure to meet scheduled production order lead times and delivery dates, "firefighting" responses, and immense sales and administration costs are examples of typical types of waste.

Accordingly, this project phase begins by including all those individuals involved in the process. Any inequality in knowledge regarding the MES project will create aggression, mistrust and arrogance and uncontrollable ancillary hierarchies will arise. Process responsibility must therefore be jointly borne by the following departments in particular: controlling, data processing, sales, production planning, purchasing, engineering, production, quality management and IT. In addition, the works council (union representatives) should already be included in this early project stage in order to counteract possible obstruction right from the start. In this connection it must be made clear that MES is not about surveillance of the workforce but is rather a way of optimizing production processes. It is also beneficial if it can be demonstrated that the workers are also direct participants in the successful implementation of the project by the fact that, say, observance of deadlines and successfully achieved quality will be positively reflected in their wage packets.

Process descriptions, such as will already have been prepared during the course of certification, are a useful resource here. In this phase the requirements of the departments or fields of responsibility involved in the process are drawn up in detail in conjunction with the employees.

3.3.3 Estimation of a return on investment

If the extent of weaknesses is documented in the form of examples and also quantified for the company, a subsequent potential analysis will help in preparing an approximate cost-benefit analysis and from this at least a rough idea of the return on investment can be calculated. This report should include not only details of the various potentials but also explanations of alternative solutions and handling approaches. It should also present the improvement potential by MES. An examination of the costs and benefits (which in this phase will still be very approximate), a first feasibility study and a rough project plan should all be constituent parts of the information provided to the persons with decision-making responsibility. The aim of this first project step is to obtain a project order including a budget and executive support from top management, ideally a "godparent" is appointed as responsible for the project.

3.3.4 System tuning

The final system tuning is based on the key data obtained from systematic process development. Requirements might, for example, include:

- Integration of the operational level (controls)
- Forms of employee remuneration
- Process time optimizations
- Implementation of quality assurance
- Interface requirements
- Special aspects of production technology
- Globalization (browser capability)
- and so on

The predefined project objectives can now be compared with MES solutions available on the market. For this purpose the VDMA offers its members, for example, specific reference lists. Considering the complexity of what MES solutions are available, it may be worth recommending getting support from a consultancy which has demonstrably proved itself to be a successful partner in the selection of this kind of corporate software. The corresponding references should be helpful here in the evaluative process. Possible suppliers will now need to be appraised on the basis of what is required.

In addition to the usual cost-benefit aspects, their offers should also be judged on the basis of "soft" factors such as:

- Financial standing. The vendor company should also supply not only documented evidence of its economic strength but also a convincing business strategy.
- The technical future potential of the solution and development strategy over the years ahead must be recognizable and transparent.
- Organizational capabilities and permanent availability are features which should not be neglected and which reveal themselves in after-sales services, software care and maintenance (update and release policies) or project know-how.

- Finally, care should also be taken that the company size of both user and supplier are comparable. It can be assumed with the mid-sized software company that it understands the daily problems experienced by the mid-sized manufacturing company. The two or three system suppliers which best meet requirements are now asked to submit their product specifications and also a detailed implementation plan. In addition, the suitability of the systems should be checked in workshops in which processes are simulated using real data from the company. At the end of this, a contract will be awarded to one of the final potential suppliers and work can start on the practical implementation.

3.3.5 Introduction of MES in the company

The implementation plan forms the foundation for the introduction of the MES system. It specifies the various stages in the introduction process and the sequence of the organizational units into which the system is to be successively introduced. It has proved helpful to introduce systems of this kind gradually and not in one fell swoop. Here it makes sense, corresponding to the 80:20 rule, to begin where the greatest potential for rationalization is located and where there is the greatest possibility of including high-tech machines and installations. Rapid, conspicuous successes are good for the image of the project and give the impetus needed for full-coverage implementation. To prevent set-backs from occurring, the real-time operation of the system must be thoroughly prepared in advance. In addition to the necessary user training, what is of prime importance here is the seamless and consistent availability of the data which the MES system needs. Here, too, the old rule still applies: garbage in, garbage out. Problems which occur during the introduction phase must be documented and precisely investigated. The actions taken to correct faults must be anchored in the introduction process in such a way that they can also be used for future fault prevention. Throughout the introduction phase, compliance with the implementation plan must be monitored not only as regards scheduled milestones but also as regards deviations from target and corrected where necessary. Right up until the end of the introduction phase, reports will be prepared at defined intervals regarding project progress and submitted to those responsible in general management. Once the introduction program has been completed in accordance with the implementation plan, the project will be formally concluded. Among other things, this means that the costs arising in subsequent periods will be clearly fenced off and that, for example, further services on the part of the supplier will either concern new projects or should be posted as recurring costs.

3.3.6 Operation of the MES solution

Added-value processes in companies are in all cases also characterized by change. For this reason it is only natural that deviations from objectives may occur even after the introduction of an MES system. On the one hand, different behaviors or processes may "sneak in" which conflict with the original objectives or which support them less. On the other hand, however, even the tactical or indeed the strategic objectives of corporate management may shift, as well as the systems supporting the achievement of these objectives. In this regard the MES system, even after its successful introduction, will remain subject to monitoring and auditing by those responsible in the controlling and organization departments. The phase when the MES solution is in operation is, however, also the time when the cost-benefit analysis is finally finished and conclusions must be drawn regarding the economic success of the project. This will be based on the same key data and measured quantities as were used as a basis for the requirements analysis. In this way management obtains not only information about the success of the project but also about the company's project capability as well as pointers regarding possible necessary improvement measures.

3.4 Innovative technologies in the MES environment

Planning and introducing the MES in order to improve information processing in production is not a one-off process but rather a decisive step towards a successful future. Here, however, the MES should not be taken in isolation but other fields of software technology should also be considered which include the factory as an information-processing system either directly or indirectly. The current and also the future role of the MES must be clearly defined if the right directions are to be set for the future.

3.4.1 The digitized factory

In recent years a widespread trend towards going digital has been noticeable in classic technologies. Familiar examples of this include digital television, digital telecommunications and digital photography. What all of these developments have in common is that today digital technologies are being used instead of analog signal processing and transmission. On the basis of modern standardized information and communications technology it is possible on the one hand to develop more powerful and more flexible products offering innovative services but also on the other hand to reduce the incidence of media discontinuities and technology-related incompatibilities.

The trend towards digital products has also long been underway in automation technology and in the production and logistics systems based on this. More and more often hardware is being replaced by software. What is of decisive importance is that even in this sector there is increasing reliance on standard IT and software. For example, proprietary field bus systems are being replaced by Industrial Ethernet, programmable controllers by soft PLCs, and special operator panels by PCs with standard Windows operating systems. In addition, in the future bar codes are to be replaced by intelligent labels (RFID chips) and cable-based communications networks by wireless technologies.

In other words, digital products and systems are spreading not only in the consumer environment but also in manufacturing areas. The products and production resources to be managed or controlled in the factory are becoming more and more communicative and mutually compatible. At the same time, however, the abundance of information and information paths to be managed by means of suitable IT and software systems is growing. In this context MES could perfectly well take on the role of a "backbone" for the factory, acting as a connecting element between the factory and the rest of the IT world.

3.4.2 The digital factory

With the digital factory, the main point of interest is not the actual working manufacturing plant but rather the ability of production to change and its flexibility. It is not only full-scale production which is continually faced with the task of replanning factories in response to new products, means of production or bottlenecks or of optimizing individual production processes. The digital factory now aims at securing an entirely digital planning process for production and factory including methodical and computer-based support. In this connection a large number of software tools have now appeared which are used primarily in production planning and factory design. The software tools create images of the machines and products together with their complete structures, logistical sequences and technological processes, going down into the smallest detail. They are then tested on a virtual basis and where applicable improved. What emerges is a well-grounded digital model of the factory, before it is even built or converted.

Today the digital factory offers the greatest potential for companies with long planning cycles or complex production processes. It is therefore no wonder that automotive manufacturers in particular are massively pushing forward the deployment of these technologies and are almost in a state of competition in their implementation. No one disputes the advantages of the digital factory for the planning process as regards shortening the start-up period, reducing planning errors and bringing down planning costs. In the meantime it is however becoming more and more clear that there is still a considerable potential for rationalization and optimization in the linking together of the digital factory and the physical factory. Feeding back real data from the factory into the digital factory has made it possible to successively improve the underlying planning models and have them reflect reality more closely. It is entirely conceivable that the digital factory will in future not only support sporadic planning processes but also be used as a tool for carrying out permanent optimization of operational processes in the factory.

3.4.3 The factory with real-time capability

Today, if the situation as regards information processing in the factory is examined as a whole, this will reveal the existence of various systems and information flows:

- Products to be manufactured are designed using CAD and are then manufactured and assembled on the basis of NC programs and drawings derived from this.
- Orders and operation sheets are planned with ERP and then finely controlled operationally via MES.
- Factories or production processes are simulated using the tools in the digital factory and then transferred into operational practice.
- Machine data are collected in the machines and installations and reported back via monitor screen or networks.

There is a general problem in the fact that these information processes are still encumbered with a large number of media discontinuities and interfaces. It thus proves extraordinarily difficult to react quickly to changing background conditions and task demands in the factory. In the light of this, research is already tackling the more advanced concept of the "real-time-capable factory" whereby the physical factory and its information and communication processes are given the ability or the authorization to react in real time to changed general conditions, faults and the like. Some approaches to implementing the real-time capable factory as regards information and communication processes include for example:

- Real-time acquisition of production or logistics data via passive or active RFID systems
- Real-time process control of machines and installations via Industrial Ethernet

- Real-time data are reported with minimum time lag for updating ERP planning
- Online provision of collected real-time data for simulation purposes in the digital factory
- Direct processing of 3D product data in production equipment such as controllers (technology-dependent)

Even if these approaches are still a long way from reality and may possibly be discarded as wishful thinking, they do nevertheless show what problems and what goals still have to be dealt with. The basic technologies for implementing these possible solutions in many cases already exist. Furthermore, intensive work is going on to get these technologies to function even better together and to reduce existing media discontinuities and incompatibilities. Further development of the system landscape and its practical use will however depend to a considerable extent on how prepared companies are to invest – even more strongly than in the past – in efficient and flexible information processing in production.

We are firmly convinced that even today a systematic use of software in production can deliver considerable added value as regards transparency, flexibility and productivity. Opening up this potential would greatly profit not only the manufacturing companies but also Germany itself as a production location.

4 MES: the new class of IT applications

4.1 Introduction and motivation

It was explained in Chap. 2 why real-time information is so important to those operational departments working close to the manufacturing event, such as production planning, production control, maintenance, quality assurance and especially the foremen. But if we look at the actual situation in the companies, it will be noticed in many cases that even today, in an age of modern, IT-supported systems, the procurement of information still leads a shadowy existence. Employees working in the above-mentioned departments are, if need be, provided with information from isolated island solutions, if at all. To get the necessary "all-round view" of all resources involved in production, the information must be brought together and supplemented by manually collected data. For today's production processes with their short control cycles the consequences are fatal: information is often incomplete or even erroneous, it comes too late and the action which is taken is often based on hunches or guesswork – in other words, on findings which are not irreproachably grounded. To close this gap is one of the most important tasks of MES systems.

However, the acquisition and evaluation of data only covers one direction of action. Modern management approaches assume that the relevant information must also be available at the machines, installations and workplaces – in other words, be directly accessible to the workers. Without being kept comprehensively informed, the employee will not be able to do his job in production to the benefit of the company. Today this means not only supplying him with the correct work instructions or inspection instructions or with drawings in their latest versions but also with further information regarding, for example, problems previously encountered in manufacturing the same part. In line with the group working concept it is in addition absolutely essential that the workers also be allowed to view the relevant key data (utilization ratios, reject rates, bonus criteria, and so on) for the production results of the work group, of the foreman's sphere of responsibility or of the department, and on this basis be able to collaborate in improving the production results.

As before, in many production facilities today data are still sent to production by the conventional, paper-based method. Time tickets and production papers are printed out directly from the ERP/PPS system, sorted manually and distributed to the relevant foreman's offices. This was not particularly disadvantageous in its effect in those days when production planning was characterized by fairly long change cycles. Today there is in many cases a time lag of several days between printing the papers for a job and when work on it actually starts and this is increasingly becoming a problem: short-term reactions to the changes (target dates, delivery quantity, quality, and so on) which today's customers demand of their suppliers as a matter of course have to be updated by hand on the already printed paperwork or the documents have to be entirely redone. The resulting problems which arise should be familiar enough, not to speak of the enormous organizational outlay.

With its systematic orientation to the requirements of those employees working in the production front line, with the MES systems there arises a new class of IT-supported applications which go beyond the classic approaches of ERP or PPS systems or of automation technology, which permit new ways of looking at things and which provide a tool tailored to the actual practice experienced by workers, foremen, maintenance engineers, production schedulers and QS officers when they carry out their daily duties.

4.2 The current situation in the manufacturing company

In the following sections we shall be describing the classic methods of information procurement and of production control and also the problems which could arise for the above-mentioned group of individuals.

4.2.1 Tools and systems for the operative level

ERP/PPS systems

A consequence of the fact that manual entries on time tickets and routing cards only become available in the ERP/PPS system with a time lag frequently of several days is that the following rule will necessarily apply: the closer to real time the information must be, the less the ERP/PPS system represents a genuine tool for the *control* of production. Information required often at very short notice regarding order progress – and thus the delivery date as well – will, for example, not usually be available until the entire order has been completed. Due to the fact that the information is posted retroactively, information about material and inventory stocks (raw materials, semi-finished products, finished articles) will not be up to date and in many cases the employees will be laboring under false assumptions. The necessary consequence is problems with the delivery dates and excessive or insufficient inventories. In addition, evaluations of orders and articles (including statistical cost accounting) are often carried out solely with regard to commercial aspects and are only available with the lack of certainty associated with a manual recording process subject to human error.

If the ERP/PPS system has to be used not only for its actual job of serving as a rough planning instrument but also as a tool for production control, problems will arise in the detailed planning. Even today a large number of ERP/PPS systems are still planning on the basis of an infinitely available capacity in production equipment. In practice there are however limitations in availability due to empty shifts and unpredictable circumstances, such as machine faults, insufficient personnel with adequate qualifications, unsatisfactory quality of raw materials, lack of corresponding tools, and so on. This lack of real-time status messages about these events, the lack of a suitable systematic facility to make this possible, and the lack of control mechanisms in the ERP/PPS system, all make the production control loop simply too sluggish. This in turn rules out a timely control response to counteract bottlenecks and situations of conflict. Production progress does not run as planned and delivery date statements made earlier bear no relation to reality.

Automation technology

The primary task of automation equipment is to regulate or control installations, processes and machines on the technical level. During recent years systems of this kind have however developed more and more into an information medium as well. As a rule it is not a problem for them to acquire, store and evaluate process values or other technical data such as, for example, machine faults. Where a shortfall can be detected when production processes are considered in their entirety is that they have no relationship to planning or logistical data. Although the foreman or production scheduler can read off the technical state of the machine directly, the relationship to the corresponding order whose manufacturing progress is affected by the fault has to be established manually in a roundabout fashion. Nor is the machine control unit in most cases aware of other resources involved in production, such as tools, the material or even the personnel.

Quality systems

Logically enough, quality assurance systems focus on all events which have to do with the subject of quality. Although there is in fact a direct relationship between the production process and the quality produced, QA systems often operate autonomously and without being integrated into production. There is no system-based connection between the production order and the inspection order and the result of this is that there is no direct planning of inspection orders or inspection, measuring and test equipment. In addition, quality data is analyzed without the necessary link to the underlying causes of quality problems on the shop floor (non-conformances caused by personnel error, poor material, tools, standstills and so on).

Usage recorders for logging machine data

The situation with these devices is similar to that with automation equipment: they have a one-sidedly technological "attitude" and the relationship to orders, to operating personnel or to the tools cannot be established, or only indirectly. Before it is possible to obtain electronic evaluations, the data have to be read out and then input manually into a separate system. Another disadvantage is the recording process which is complex, requires intensive servicing and is also expensive (consumption of special forms, ink cartridges, and so on), not to mention the entire organization which has to put it place for supplying the consumables. Even the electronic pendants of the usage recorders (MDC systems, for example) will not improve the situation unless an automatic link is set up to the world of the orders.

Obsolete PDA systems

In the case of the PDA systems which were introduced a few years ago and which are subject to IT-related limits, in most cases users took a special approach. They were, for example, used as an information source for the data which are created at focal machines or operations, or when orders are reported as completed. In most cases they can only cope with evaluating single-stage production processes (for example, when the focal operation is injection molding) and are often only island solutions which lack interfaces with the ERP/PPS system. In addition, here too a field of view covering all resources is lacking, for example, covering tools, personnel or materials.

4.2.2 Manual information procurement and other tools

To find a way out of the deficiencies described above, many companies are setting up facilities in addition to the systems mentioned with whose help urgently required information is obtained about the current event or planning activities supported.

Planning tables or pinboards

The advantage that the frequently used planning table or pinboard has over PC monitors is that it provides a clear large-area display of the planning scenario. Its function is, however, based on printouts of order documents from the ERP/PPS system and their complicated, manual deployment on the pinboard. This means a lot of effort required when new orders are added and especially when already inserted order cards have to be rescheduled (urgent orders with a high priority, change in priorities, unplanned delays). Apart from visualization, the pinboard offers no kind of support for availability checks, for reviewing concatenated events or even in ascertaining what is the real capacity available, limited as it is by machine faults or lack of personnel.

Progress chasers

As has been stated in the preceding chapters, without a PDA or MES system no current information about order progress will be available and for this reason no precise statements can be made "at the touch of a button" about delivery dates and quantities. But since customers demand concrete information, in many companies the position of "progress chaser" has been created. His task is to gather all the necessary information about customer orders by walking the shop floor up and down. This involves a lot of personnel time commitment and also means time delays caused, for example, by searching for parts.

Time tickets and routing cards

These are used for "transporting" the work on hand into the production areas and making information available about the orders or operations. They do not, however, always reflect the current status since they are printed out from the ERP/PPS system and do not automatically keep up with quantities changed at short notice, with a new machine allocation or with changes in operation sheets and inspection and test plans. A lot of labor is involved in the workers filling out their time tickets and no plausibility check is made of the data. Further expense, time delays and often even errors arise since other employees then have to input the data by hand into the ERP/PPS system.

Clock cards

Even today clock cards are still used in many companies for documenting the employees' clocking-in and clocking-out times and from these working out the hours they have worked as a basis for wage calculations. This involves a great deal of outlay on administration and materials. Here too additional labor capacities are required to calculate times and to input them into the wages and salaries system. In addition, an overview of which employees are present and absent is only available locally at the time clock.

Work instructions, drawings and inspection and test plans

Printed information is also a medium to which workers, line engineers, inspectors and foremen are accustomed. However, even these documents do not always show the current status since printing from the ERP/PPS, CAD or QA system does not take place in real time when production starts. A huge organizational and administrative overhead is involved in preparing, updating or distributing the documents.

4.2.3 Problems in bringing together the data

The systems and tools described above are in most cases island solutions and offer no way of exchanging data with other systems or if so only to a limited extent. The necessary all-round view is not however possible without an exchange of data or a comparison of data. Let us provide a few examples:

- To ensure gapless registration of productive times, the employees' attendance times are compared with the productive times from PDA. This comparison is absolutely essential if performance-based or incentive payments are made.
- In short-term manpower planning the foreman on the one hand needs the information from time-off planning and staff work time logging (who is actually at work?) . On the other hand, the job load which has to be managed must of course be available as a result of detailed planning in order to make it possible to determine personnel requirements.
- Data must be brought together from different areas and systems when an explicit POP (proof of product) is required and when it must be documented which employee made which part using which tool on which machine under what process conditions and using which batch of raw materials.

4.3 The situation as it ought to be

Looking at the actual situation we have described, it very rapidly becomes clear that procurement of information, the information flow and detailed planning in the manufacturing company have to be significantly improved.

Which general conditions will have to be observed and what benefits an MES system can offer will be dealt with in the following sections with the aid of a number of representative examples.

4.3.1 Gapless automated data acquisition

The road to networked information in production commences by collecting data from all relevant resources and processes as gaplessly as possible and storing them in an all-encompassing database. In contrast to the traditional recording method, with an MES the value-adding process are recorded online and “without

With their holistic approach MES systems will deliver a further jump forward if not only the productive times are registered but also those ancillary and ineffective times which today are typically ignored (burden rates, waiting, transportation or queuing times) as well as idle times and downtimes. This automatically produces a better database for statistical cost accounting and for tracking down the real origins of costs.

It should however be noted that the completeness of the data is directly related to the effort put into acquiring the data. MES systems will cause acceptance problems if the additional load on the employees no longer stands in reasonable relationship to the beneficial effects which can be achieved. For this reason simplifications should be made here in a form whereby the data are, for example, automatically acquired via a direct connection with the machines, weighing devices and other equipment, and manual data input is eschewed as much as possible. The use of machine-readable identification bearers (such as bar-codes, transponders, RFID

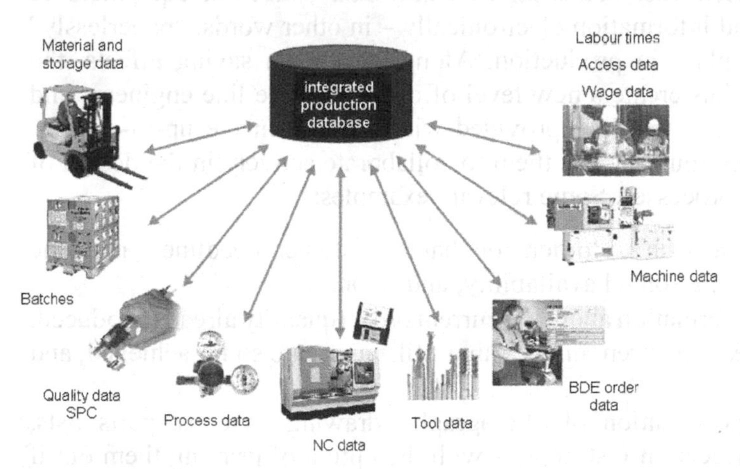

Fig. 4.1. The integrated production database as a necessary condition of the “all-round view” in production

tags, and so on) for transferring the saved data renders superfluous the expensive and potentially error-riddled method of typing the data in manually. Mature MES systems also offer additional ways of making things easier such as data acquisition logic units which considerably reduce the time and effort involved in inputting data. Some relevant examples:

- In the case of orders whose processing times extend over several days, when the employee clocks off at the end of his shift he "pauses" his current job only to resume it automatically when he clocks back on again the next morning. In this way the order no longer needs to be repeatedly suspended and resumed manually.
- The time difference between the completion signal for an operation forming part of a multistage order and when the successor is logged on is automatically interpreted by the MES as transfer time or burden time. Separate postings are not required here.
- Several events of short processing duration are simultaneously logged on and off at a PDA terminal as so-called collective operations. This disposes of problematic inputting and travel times. The MES automatically posts a proportional time for the individual events on the basis of configurable rules.

4.3.2 The information point for production

The introduction of production-proximate systems is usually also bound up with the installation of an IT infrastructure which extends as far as the work centers and machines. Powerful MES systems also have the ability to use existing networks, industrial PCs and data collection equipment to convey data and information electronically – in other words, "paperlessly" – to the right place in production. Alongside known saving effects, the MES systems thus create a new level of quality for the line engineers and machine operators: they are provided with comprehensive up-to-date information which thus enables them to collaborate actively in the design of the production processes. Some relevant examples:

- Planning data such as orders on hand, customer deadlines, machine maintenance, personnel availability, and so on.
- Display of information about the current order (quantity already produced, calculated lead time remaining, labor utilization rate so far achieved, and so on).
- On-screen presentation of photographs, drawings, videos, parts lists, work and inspection instructions with the option of printing them out if necessary.

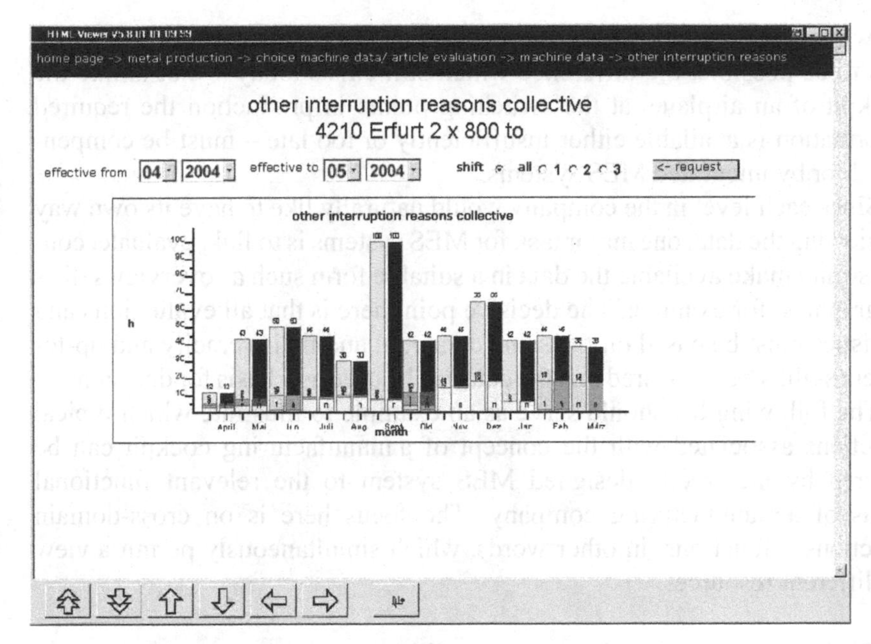

Fig. 4.2. Evaluation of machine idle times

- Performance comparisons and statistics regarding reject rates, deadline delays, key data, reasons for idle times and downtimes etc. so as to be able to estimate the status of individuals, of the department or of the entire plant.
- Current target/actual comparison for key data such as machine utilization ratio or order fill rate and machine performance or cycle.

Personal information about remaining vacation, overtime or flextime accounts or about the performance rate achieved in the case of bonus or incentive pay. Figure 4.2 shows by way of example how employees are kept comprehensively informed and can thus be included in the decision-making processes: evaluative material regarding machine idle times for any selected period is available at the information point. Naturally it is also possible to print out overviews of the key data and statistics worked out in the MES and then to publish them in paper form in the production department concerned.

4.3.3 The concept of the "manufacturing cockpit"

In many companies the duties of a production scheduler, maintenance technician, quality assurance representative or foreman are the same as those of management and are comparable with the duties of a pilot. Problem situations must be detected rapidly and suitable action taken with the smallest time delay possible in order to prevent escalation from occurring.

So why shouldn't the cockpit concept also be applicable to production? For this to be possible, the difference which still exists today – that unlike the cockpit of an airplane, at the switching points in production the required information is available either insufficiently or too late – must be compensated for by integrated MES systems.

Since each level in the company would naturally like to have its own way of viewing the data, one major task for MES systems is to link, evaluate, condense and make available the data in a suitable form such as overviews, lists or graphics, for example. The decisive point here is that all evaluations and statistics must be based on the same data pool and their veracity and up-to-dateness thus be so assured that the data can be used as a basis for decisions.

The following list should serve as an example to illustrate which typical functions associated with the concept of a manufacturing cockpit can be offered by a correctly designed MES system to the relevant functional areas of a manufacturing company. The focus here is on cross-domain functions – functions, in other words, which simultaneously permit a view of different resources.

MES functions in the foreman's office

- Up-to-date overviews of orders and machines to allow rapid detection of problematic situations
- Simple planning tools to enable stipulation of sequence in which orders will be processed and to enable orders to be rescheduled

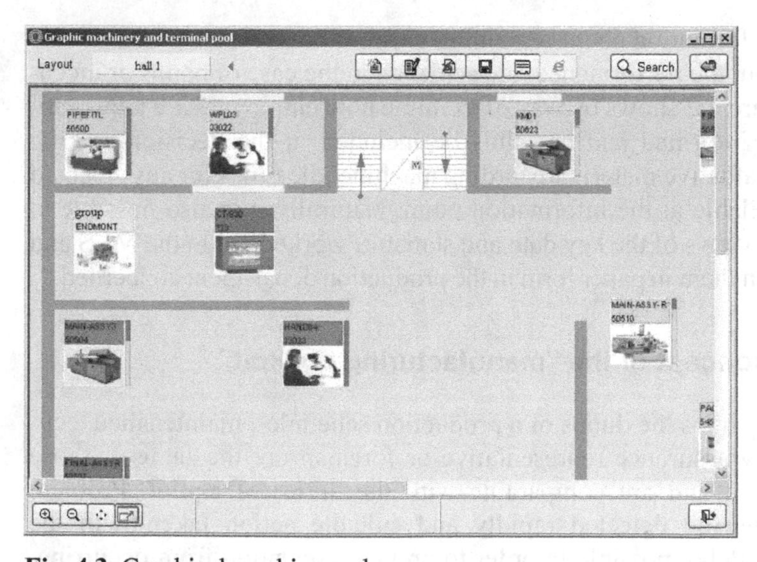

Fig. 4.3. Graphical machine park

- Planning of vacation and non-productive times for the assigned employees as part of short-term manpower planning
- Up-to-date overviews of which employees are present and which absent
- Order- and personnel-specific shift records
- Idle-time evaluations for machines and installations
- Overviews of the qualities currently being produced

With the graphical machine park shown in Fig. 4.3 the foreman has at all times an overview of the current status of his machines and orders. The decisive advantage here is that all information is available at the same time, thus enabling the person in charge to react immediately. In this way a time-lagged statistics function is turned into an active closed-loop control function.

Operations planning and production control

- Multi-order analyses of production progress including extrapolation functions and automatic planning aids
- Lists of materials to be processed and retooling lists
- Complex detailed planning tools on the basis of graphical planning tables
- Order and article statistics permitting inferences to be drawn as to "how things went" for an identical part in previous orders
- Availability analyses and checks for machines, tools, personnel and materials

The graphical planning table is the central information and detailed planning instrument in production control. Capacities are displayed in the table

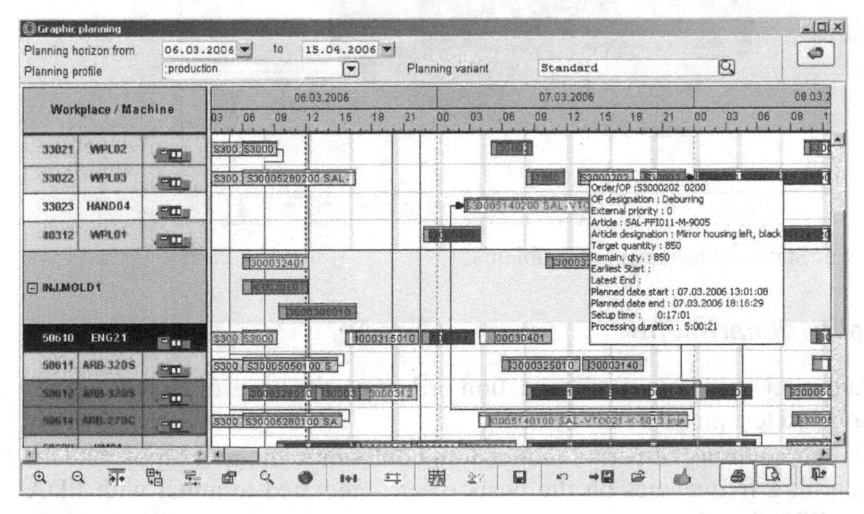

Fig. 4.4. Graphical planning table

for both their current and also their future states, with no need for data entry. This means that imminent conflicts can be detected in advance and corrected.

Maintenance

- Graphical machine park with an up-to-date display of machine statuses (for example, even including a projector in the shop)
- Maintenance calendar for machines and tools
- Statistics for cause of problems, with configurable levels of detail
- Fault class evaluations
- Graphic plots (profiles) of key data important to maintenance (curve plotted for degree of utilization, machine cycle, and so on)

Statistics relating to problem causes and idle-time statistics as shown in Fig. 4.5 make it possible for the person in charge (group, foreman, maintenance engineer, and so on) to specifically track down what is causing problems in machines and installations.

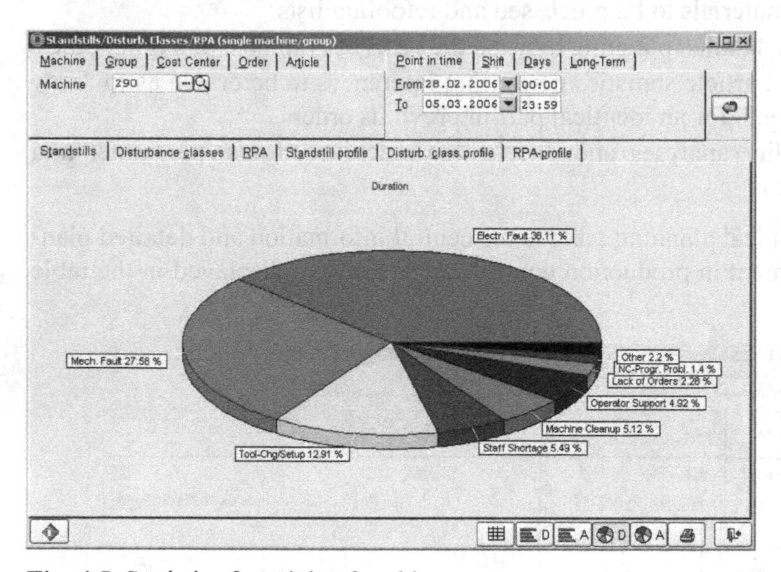

Fig. 4.5. Statistics for origin of problems

Quality department

- Automatic generation of inspection orders on the basis of stored inspection and test plans
- Online counting of parts produced and automatic monitoring of random sampling frequencies on the basis of an integrated terminal with PDA and SPC functions

- Production control station with facility for checking the availability of inspection and test plans
- Registration of batch and lot information within the normal PDA entries
- Generation of product history record for intermediate and finished products or of a where-used list for raw materials and semi-finished products

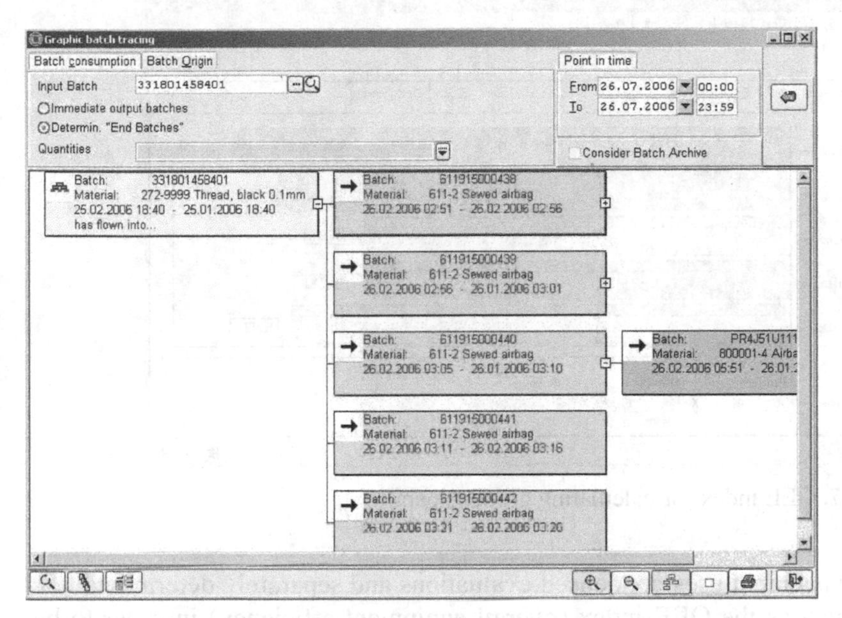

Fig. 4.6. Batch and lot tracking

If batch and lot information is also registered in parallel with PDA messages, a product history record (batch tree) will be automatically created for the manufacturing events. Batch and lot documentation is becoming more and more important, particularly in the light of the requirements of modern certification rules (ISO 9001/TS 16949). In general it is true to say that a very large number of auditing requirements cannot be cost-effectively solved today without an integrated MES system.

Controlling and management

- Evaluations of the percentage utilization of machines and of free production capacities in short- and medium-term planning periods.
- Support for the continuous improvement process by means of long-term observation of changes in utilization rates and other key data.
- Use of manufacturing scorecard methods (MSC) (Kletti and Brauckmann 2004) for the continuous monitoring of defined objectives.

– Condensed evaluations and statistics relating to "problem areas" in the company such as sickness levels, delivery problems, failure to meet deadlines, trends in scrap rates, problem originators (machines, tools, produce parts), trends in queuing times, and so on.

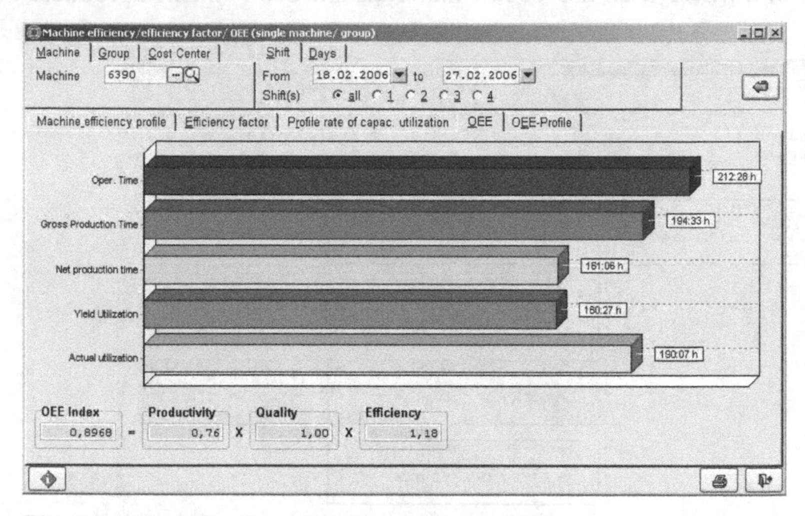

Fig. 4.7. OEE index for calculating group capability

Management uses condensed evaluations and separately determined key data such as the OEE index (overall equipment efficiency) in order to be able to precisely evaluate the efficiency of machines and installations. Objectively measurable key data are thereby made available to the operational level for the first time and thus become the foundations for a process of continuous improvements. In the past it was not possible in particular to solve the problem of process-relevant objective key data on the basis of accounting department capabilities.

Personnel department

– Up-to-date attendance and absence lists.
– Delegation of time-consuming routine activities (approvals of vacation applications, reception of notifications of sickness, planning relating to non-productive time) to the foreman and thus to the place where the information usually comes in directly or where it needs to be processed anyway.
– Comparison lists which provide an automatic comparison of attendance times and productive times.

- Automated calculation of performance-based or incentive payments and the necessary automated integration in wage and salary systems.
- Setting up a personnel information system with data which would, for example, be useful in short-term manpower planning.

The shift plan in Fig. 4.8 will display at the touch of a button how many and which employees are available on which day for which shift, and which non-productive times are planned in (due to vacation or further training, for example).

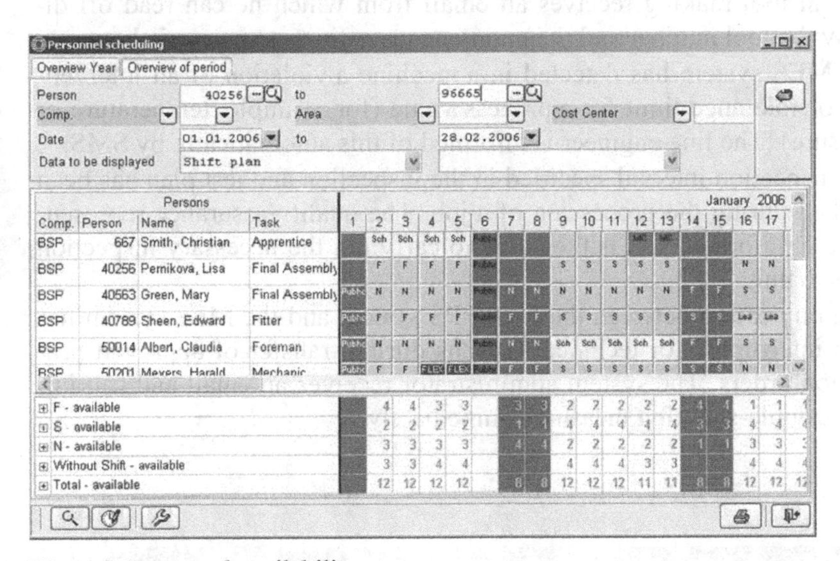

Fig. 4.8. Personnel availability

4.3.4 Escalation management and workflow

What virtually all of the examples we provided above have in common is that the data are called for as needed by the corresponding organizational units. An entirely new quality of information provision is offered by MES systems which include an integrated escalation management function and an individually definable workflow process based on this. The decisive advantage lies in the required information being directed automatically to the correct location and the person in charge no longer having to actively trouble himself with doing this. This means we have real-time notification when a critical situation arises and it is necessary to intervene in processes which are running.

In addition, a workflow can be created for each escalation or for each event. Within this workflow it is specified how notification will be implemented (for example, SMS on cell phones, by email, as a pager message, or as a pop-up window on the PC screen). If the notification is not acknowledged by the receiver within a specified time, the next escalation level will ensure that the message is also sent to his deputy or his supervisors. A number of examples will now follow which are intended to illustrate practical applicability in the manufacturing environment:

- The maintenance technician receives a message on his pager telling him that a particular type of fault has occurred in a machine.
- A tool has become due for scheduled maintenance. The person responsible in tool making receives an email from which he can read off directly the tool number and the maintenance activity to be carried out.
- The MES system has detected in a machine a violation of an intervention or tolerance limit for a process value (for example, temperature or pressure). The line engineer is informed of this automatically by SMS.
- The inspection interval specified in the inspection and test plan has been reached during the processing of a job. The quality assurance representative is informed and is then able to carry out the necessary inspection immediately.
- Communication between the ERP/PPS system and the MES system has been interrupted for technical reasons during transfer of approved production orders. The system administrator receives an email and can initiate trouble-shooting measures immediately.

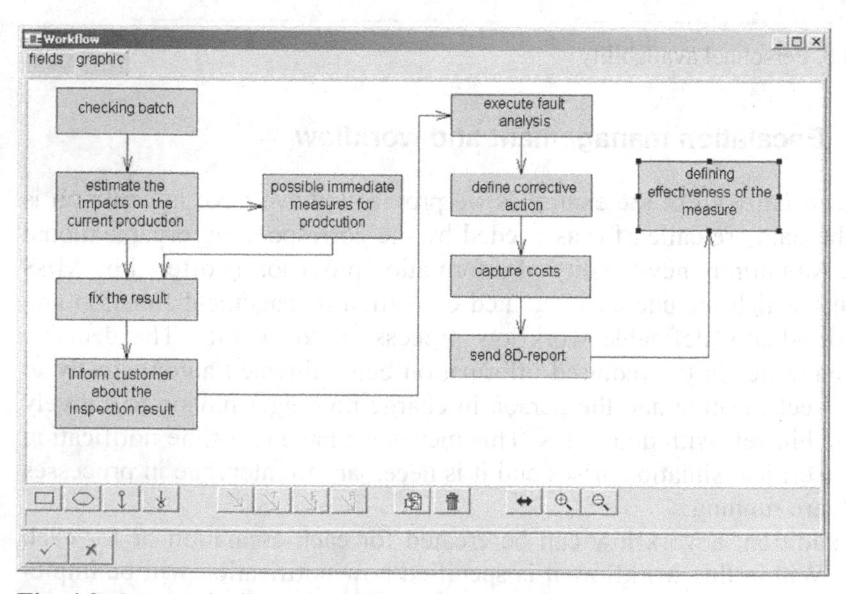

Fig. 4.9. Automatically generated workflow for error processing

An example of a workflow. As soon as a certain value is crossed a workflow (also predefined) can be generated by the MES which specifies the involved employees to process the error. All processing steps are written to a response tracking file with the date and the name of the employee responsible.

The screen shot shows an overview of all escalations occurring within a selected time period. The data evaluated enable conclusions to be drawn regarding the reaction time, persons involved in processing the problem, and termination of escalation.

Other typical application cases can also be found in the personnel and security fields. For example, a workflow can be set up for when an employee submits his vacation application at an information point (PC or terminal with web browser) and his supervisor is automatically notified about this. Following approval or rejection of the application the employee will receive the corresponding message at the time registration terminal when he clocks out in the evening.

4.4 Outlook and further development of MES systems

To the same extent as manufacturing companies will in coming years need to meet the changing demands of the market, MES systems will become more and more important. Short delivery times, high upward pressure on costs, smaller batch sizes and stricter quality requirements call for a highly

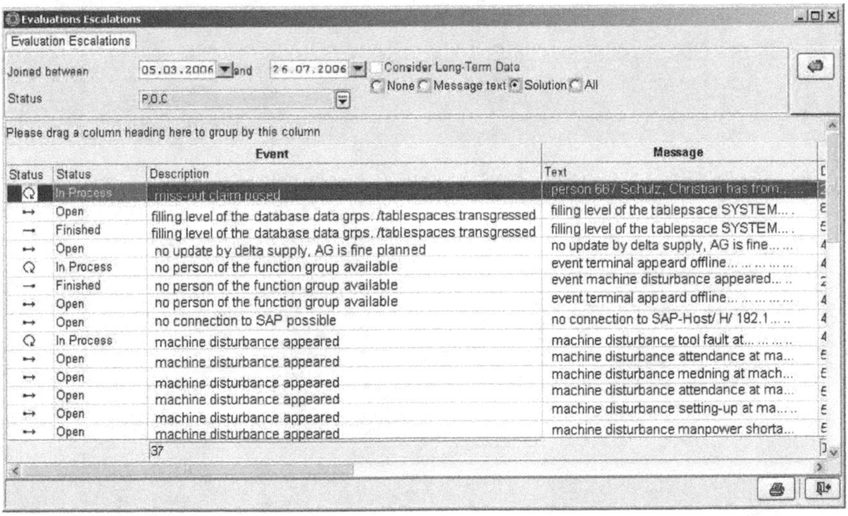

Evaluations Escalations

Evaluation Escalations

Joined between 05.03.2006 and 26.07.2006 Consider Long-Term Data

None Message text Solution All

Status P,O,C

Please drag a column heading here to group by this column

Event			Message
Status	Status	Description	Text
	In Process	miss-out claim posed	person 66 / Schulz, Christian has from...
	Open	filling level of the database data grps. /tablespaces transgressed	filling level of the tablepsace SYSTEM... .
	Finished	filling level of the database data grps. /tablespaces transgressed	filling level of the tablepsace SYSTEM... .
	Open	no update by delta supply, AG is fine planned	no update by delta supply, AG is fine... ...
	In Process	no person of the function group available	event terminal appeard offline... ... ...
	Finished	no person of the function group available	event machine disturbance appeared... ..
	Open	no person of the function group available	event terminal appeard offline... ... ...
	Open	no connection to SAP possible	no connection to SAP-Host/ H/ 192.1... ..
	In Process	machine disturbance appeared	machine disturbance tool fault at... ... .
	Open	machine disturbance appeared	machine disturbance attendance at ma...
	Open	machine disturbance appeared	machine disturbance medning at mach...
	Open	machine disturbance appeared	machine disturbance attendance at ma...
	Open	machine disturbance appeared	machine disturbance setting-up at ma... ..
	Open	machine disturbance appeared	machine disturbance manpower shorta...
		37	

Fig. 4.10. Tracking corrective action

process capable production organization which is able to react flexibly to the needs of customers and to internal constraints – and do so without losses in quality or increased costs having to be accepted. Only by using integrated MES systems will it therefore be possible to reduce internal friction losses, to produce less expensively and to have better control over the production processes. In the light of this the MES functions we have described will in future be an indispensable tool for all company departments in handling their daily tasks.

Literature

Kletti J, Brauckmann O (2004) Manufacturing Scorecard – Prozesse effizienter gestalten, mehr Kundennähe erreichen – mit vielen Praxisbeispielen [The manufacturing scorecard: designing processes more efficiently, achieving greater customer orientation, with many practical examples], Gabler, Wiesbaden

5 Building an MES system

The demands made of a modern MES system mean that these systems must be given an appropriate structure. Island solutions or even classic PDA systems are biased towards being able to run several monolithic software modules in parallel on one integration platform. Users have always wanted these monolithic modules to be able to communicate with each other. Unfortunately the reality was often different, such as, for example, the fact that in one earlier PDA system the shift performance of the machine and the number of parts produced for that operation had to be input in two separate dialogs at the end of the shift.

One of the most important reasons for these problems was the demand of the market for a standard software system. Instead of requiring individual programming and thereby making it necessary to carry out a corresponding analysis of requirements, these standard products promised the user a fast and inexpensive way to reach his goal. The limits of these solutions are easy to see from the example cited: mutual integration of these products only occurred at precisely the place where it was explicitly planned. Should this integration not have originally been deemed necessary, it would not then be available when actually needed and would thus have to be purchased later on, and in some cases at considerable expense.

In recent years manufacturers of MES solutions have been engaged very intensively on breaking through the limits of existing software architectures. The keywords "process mapping", "business logic" and "process workflow" have in the meantime become familiar to every decision-maker and to every consultant in the MES field. Here an analysis of the concepts very rapidly reveals that in many cases they are understood differently. In fact behind the term MES and the corresponding concept is a modern software architecture which should basically cover the following requirements:

- Complete mapping of all requirements beneath an ERP/PPS system (so-called horizontal integration)
- Availability as standard software with the following properties:
  - Modular software structure
  - Expandable in accordance with the requirements of the user based on current standards

- Simple adaptability of the standard modules not only to the processes but also to the functional requirements of the user
- Availability of standardized interfaces on all levels

The first two points are a basic requirement of any MES system which is to be taken seriously. Without complete coverage of all the user's requirements under an ERP/PPS system the user will be purchasing precisely the problems he is actually trying to avoid. The standard software mentioned in the second point above must, as will become clear later, be available in conformity with the requirements of the MES architecture and will therefore differ from the old monolithic standard products mentioned at the beginning. Only then will it be possible to fully exploit the advantages relating to adaptability to the user's processes. With the fourth point, a flexibility is provided which marks out an MES system as an open and expandable system. The last two points together represent the basis by which an MES system will in future be simply and flexibly adaptable to the requirements of the user regarding the mapping of his changing processes.

This chapter describes the architecture and the structure of an MES system together with the components it needs. The reader will thus gain an overview of the individual constituent parts of an MES system. It should be possible on the basis of this information to select and evaluate MES systems by their operative capabilities and flexibility. This chapter will also seek to provide information on how a modern MES system can handle the variable mapping of processes which is so important to the modern company.

5.1 Software architecture of an MES system

The architecture of modern MES systems is also oriented like other business solutions by the so-called "business service architecture" or "enterprise service architecture" (ESA for short). An important reason for selecting this architecture is that during the life cycle of an MES system new demands are continually arising which need to be implemented in a virtually never ending process.

The individual layers of the architecture of a modern MES system and their special features are explained in the following sections.

In addition to ensuring the delivery of predominantly technical properties, both the architecture and the basic functions of an MES system have a special importance. Provided that both the architecture and also the basic functions are properly defined and implemented and the basic functions are oriented to the standards usual in the market, this will form the best conditions for an open, expandable and future-proof system.

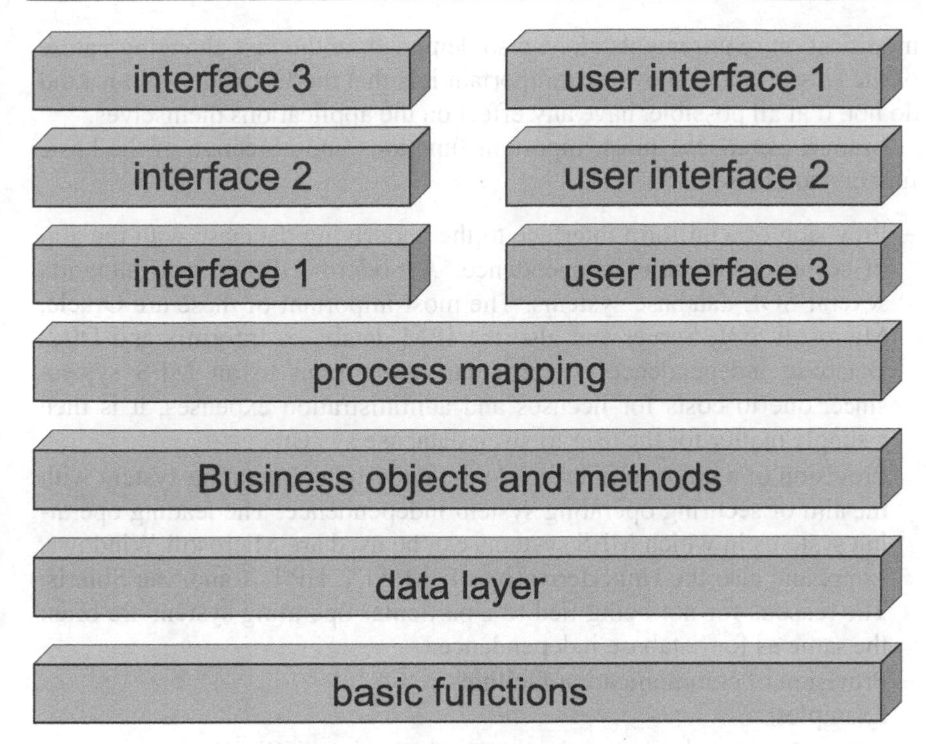

Fig. 5.1. The software architecture of an MES system

This is why many proprietary systems suffer from the lack of a global architecture or from the absence of any way of allowing new technologies to feed in rapidly and smoothly within the basic functions without unwanted effects on the applications themselves and thereby constituting an IT risk.

5.1.1 Basic functions

The basic functions of an MES system are a collection of software functions which are available independently of a particular product and which enable the products based on them to be developed as far as possible using the same modules and with a uniform structure. As has already been shown in the previous section, a good basic functionality allows technologies to be incorporated or replaced without this affecting the individual products.

The basic functions will therefore provide separation of the MES applications from the technical components which form the basis of any IT system. In practice the users, often compelled by corporate policies, want a change in the operating system, a change in the database system or a change in the communication protocols for their MES application. These modification requirements alone also demonstrate the fast-changing nature of the IT sector and show how important it is that modifications of this kind do not, if at all possible, have any effect on the applications themselves.

In more detail, the most important functions and objective of the basic function are these:

- Provision of a uniform interface to the underlying database with the aim of securing database independence. A modern MES system supports several SQL database systems. The most important of these are Oracle, Microsoft SQL Server and also the IBM databases Informix and DB2. Database independence is particularly important to an MES system since, due to costs for licenses and administration expenses, it is then a simple matter for the user to swap database systems.
- Provision of a uniform interface to the underlying operating system with the aim of securing operating system independence. The leading operating systems in which MES systems can be used are Microsoft Windows, Linux and also the Unix derivatives IBM AIX, HP UX and Sun Solaris. The reasons for not being tied to a particular operating system are often the same as for database independence.
- Provision of communication facilities
  Examples:
  - Secure network communication on the basis of TCP/IP
  - Bus systems in production
- Provision of components for typical MES tasks
  Examples:
  - Components for displaying business diagrams
  - Components for the temporary storage of data
  - Components for the secure acquisition and checking of data
- Provision of interfaces and functions for incorporating products with the components data layer, application layer, and process mapping.
- Provision of separate database partitions for OLTP[1] and long-term storage to ensure response times on the one hand as well as medium- and long-term availability of data on the other hand.
- Provision of technologies for interfaces
  Examples:
  - Web services
  - OPC
  - Excel export or XML export
  - Various file formats

---

[1] OLTP is an acronym for online transaction protocol and is used here as a synonym for the real-time database section of an MES system

- Product-independent alarm system with the corresponding communications terminal points, email, cell phone, pager and so on (so-called escalation management).
- Functions for logging, monitoring and tracing (for example, for the rapid detection and locating of fault states).

For an open system it is also important for the MES manufacturer to keep some of his interfaces open, particularly for those interfaces and functions relating to integrating products. If this is the case, then partners or even IT-savvy users will potentially be able to integrate their own applications into the existing MES system.

Especially when an MES system is used in food or pharmaceutical industry it is subject to additional technical requirements relating to so-called "FDA conformity" which will ideally already be reflected in or at least supported by the basic functions of an MES system. In addition, a supplier of suitable MES systems to the food and pharmaceutical industries will have knowledge and experience of engineering and developing software in conformity with FDA requirements.

5.1.2 Data layer

The data layer is that part of the MES application which is responsible for the definition of the database structures as well as for the data stored in them and thus looks after the so-called persistence of the data. Every MES product possesses the data model corresponding to one version of a product. Today this data model is normally stored in relational database systems and the corresponding data processed by means of SQL (structured query language).

The necessity of defining a data layer reveals itself in the overall view by the way products are mapped in the architecture of a modern MES system.

It can be seen from the diagram above that the products in an MES system are spread over three layers of the system architecture. From this model a rule can be derived in the form that changes in or expansions of a product whenever possible take place only in one layer. This rule ensures stability and reduces the effort required to implement changes. The change itself takes place precisely in the layer which is responsible for the change.

Here the data layer handles the task of ensuring that the data for an MES product based on the underlying database system can not only be written reliably and permanently but also read again. The data layer defines the necessary tables and fields in the database system. All changes, modifications and product expansions relating to the persistence of data will thus take place in this layer. Other examples of intervention in the data model

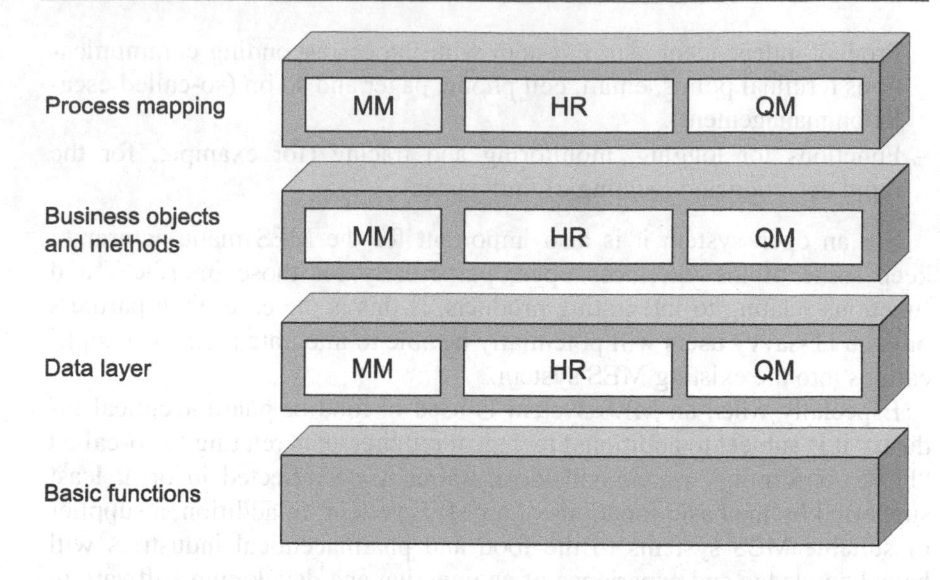

Fig. 5.2. Product image in the enterprise service architecture

are changeovers of database structures due, for example, to new functionalities or to a further system expansion into additional production areas or to cover additional applications. It is therefore clear that the data model often undergoes technical changes which conform only to a qualified extent with the requirements made of the application.

This gives the user the ability to access the data layer directly via reporting tools in order to produce his own analyses in an elegant and simple manner. Due to the changes we have described within the product, when the product is upgraded or when there are new versions of an MES product it frequently happens that the evaluation based on the data layer needs to be adapted to the modified data layer. From this it follows that direct accesses to the data layer cannot be classified as "release-proof". Compatibility with future releases will not exist unless access to the data is via the application layer.

5.1.3 Application layer: business objects and methods

The application layer depends on the data layer and makes functionality available to the process mapping. The application layer covers the following important requirements:

- The application layer provides the objects and corresponding methods for creating the business logic in the process mapping layer.

What is meant at this point by an object is, for example, "Machine 3523 with associated data" or "Operation 7330022 010 with associated data". A method means the functions by which the data of the object can be processed or functions which trigger actions for the object, such as logging someone onto a machine, for example.

- The application layer supplies its objects and methods irrespective of the data model. This procedure ensures that if there are changes in the underlying data structures the application layer will look after the necessary compatibility and the objects and methods will behave in their normal way. In the case of new functions, the application layer will explicitly make new objects and new methods available for them.

The demands made of the application layer show clearly that this layer has the important job of getting necessary technical modifications under control so that the process mapping can function as normal. Provided it has been correctly implemented, the architecture of an MES system will thus ensure that changes do not negatively impact already defined processes.

5.1.4 Process mapping

The process mapping has the job of reproducing the actual business logic on the basis of the methods and objects of the application layer. One synonym for the term business logic is enterprise logic.

Via interfaces or graphical interfaces the process mapping receives messages including the corresponding data which essentially are then processed via "if-then" conditions and the use of the objects and methods of the application layer. The process mapping thus contains the actual logic of an application or of a product and in addition makes use of the underlying layers. One important characteristic of the process mapping is that the objects and methods of all products are at its disposal. This means that several products can be "woven" together in the process mapping without any problem, something monolithic products do not permit.

This advantage becomes considerably greater if further products are to be added at a later date. If necessary, the methods of the new products will simply be incorporated into the existing process mappings. It should now be clear that these possibilities permit complete horizontal integration and interweaving of the applications, something which can also be carried out gradually.

A further advantage of the process mapping is the high degree of compatibility with future application releases which this layer provides. The underlying layers ensure compatibility in the event of revisions and modifications such as product upgrades, for example, so that the process mapping created is not affected.

For the if-then relationships, the classic configuration settings can be used as before. These are supplied to the process mapping from the application layer by means of the corresponding methods. The if-then conditions we have mentioned can, however, also be represented graphically: the result is the so-called workflows. The advantage of a graphical representation is that on the one hand these workflows offer clarity and describe process flows in a readily comprehensible manner while on the other hand there is also the possibility of creating these workflows in electronic form and then generating the process mapping from them.

5.1.5 The advantages of ESA architecture for MES systems

In systems without an architecture of this kind, changes in the processes and process sequences of the user often mean immense trouble and costs within the software being used since considerable intervention in the standard processing routines is required, particularly in the case of product-independent changes. The architecture we have described of an MES system can, when used correctly and consistently, guarantee considerable reductions in costs and a faster implementation. In the most favorable case all that is needed is an intervention in the "process mapping" – in other words, the standard products being used need no modifications.

Other advantages of an ESA architecture for MES systems:

- The limits of classic configuration methods disappear with the ESA architecture and new possibilities such as graphical workflow deliver clarity, transparency and security.
- A complete horizontal integration is realistic and expenses are much lower than before.
- It is now a good deal easier to introduce the solution gradually.
- Processes can be mapped in the form of graphical workflows.
- It is much easier to modify the process sequences than is the case with monolithic products, even after the introduction phase.
- The risk of new errors creeping into existing process sequences is reduced.
- Changing products involves less effort.
- There is a reduction in test phases and testing expenses.
- Partners or even IT-savvy customers are potentially able to create equivalent products or modules at the MES supplier.

The following diagram shows the difference in a product-independent change between the classic architecture and ESA architecture.

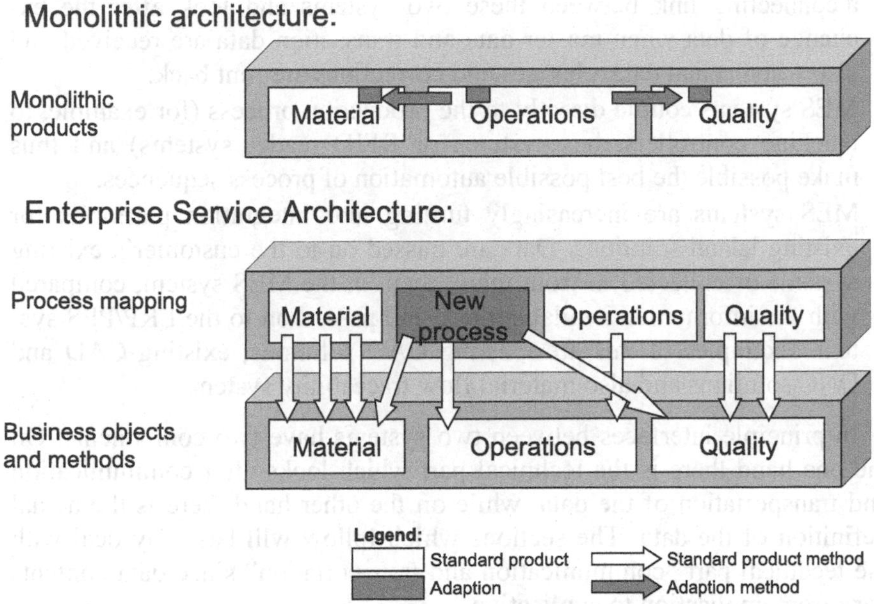

Fig. 5.3. Planning depths with monolithic architecture and with Enterprise Service Architecture

As can be seen, in monolithic architectures, interventions in several standard products are necessary and these changes must also communicate with each other correctly. Within an ESA architecture, the change can in the most favorable case be made by setting up a new process which uses existing calls to the standard products.

This method will thus be possible when the products of the MES supplier are structured in a way corresponding to the ESA architecture – in other words, in addition to the user interface and the associated interfaces these products have their own data layer, their own application layer and also their own process mapping.

5.2 Interfaces of an MES system

Interfaces means the communication devices which different systems use for exchanging data. Provision of and support for standardized interfaces is indispensable to a modern MES system for the following reasons:

- MES systems differentiate and automate the processes of ERP/PPS systems and condense the technical data in the MES environment into information suitable for these ERP/PPS systems. Interfaces function as a connecting link between these two systems and look after the exchange of data when master data and transaction data are received and also when actual data, changes and corrections are sent back.
- MES systems couple directly to the production process (for example, to machine controllers, bus systems or RFID reader systems) and thus make possible the best possible automation of process sequences.
- MES systems are increasingly turning into integration platforms for existing island solutions. Data are passed on to the customer's existing systems or collected in from there, then, in the MES system, compared with the information available there and passed on to the ERP/PPS system. Examples of this are existing CAQ solutions, existing CAD and DNC solutions and also material flow traceability systems.

In principle interfaces between two systems have two components. On the one hand there is the technical part which looks after communication and transportation of the data, while on the other hand there is the actual definition of the data. The sections which follow will basically deal with the technical part "communication and transportation" since data contents vary from application to application.

When selecting an MES system care should always be taken that the system actually supports new communication paths and the data contents will need to be defined flexibly for the interfaces. How can a state-of-the-art technology benefit the user when every little adaptation calls for time-consuming, wearisome programming? In this regard a good MES system can be recognized by the fact that the data contents of interfaces can be adapted to the requirements of the customer in the simplest way possible – as part of customizing, for example.

5.2.1 Interfaces with higher-level systems

The following sections will provide an overview of implementing the task of "communication and transportation" which must be handled by interfaces to higher-level systems such as ERP/PPS systems, wage and salary systems. In addition, brief appraisals will be given of the individual possibilities.

Interface from database to database

The question as to why database cannot simply be linked to database crops up again and again. Possible technical interfaces for this are not only the

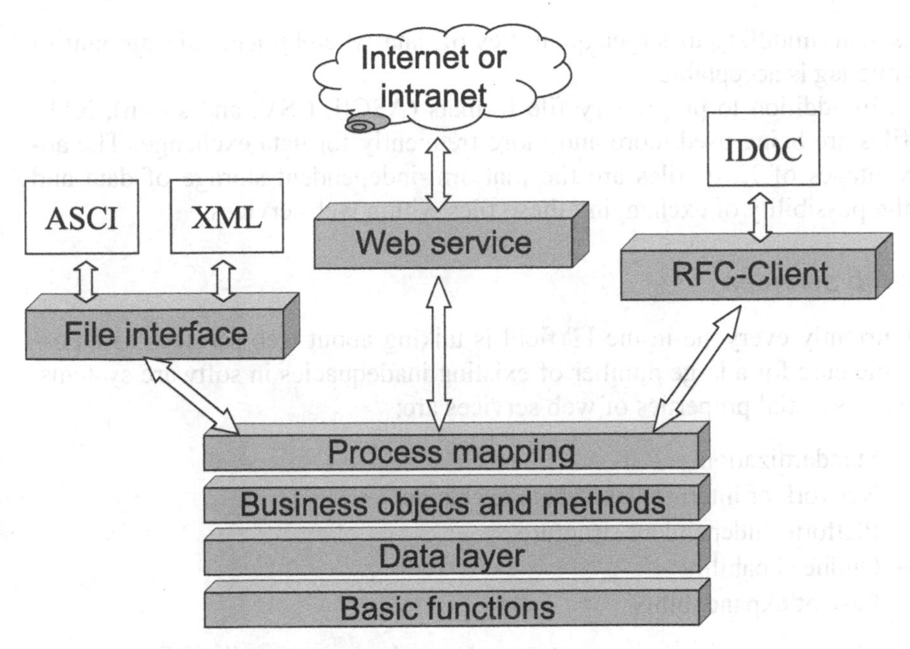

Fig. 5.4. Interfaces with higher-level systems

so-called native drivers from the database manufacturers but also ODBC or JDBC. The reasons why people wonder about this are obvious since definitions of the database structures are available and one would expect that rapid implementation would be possible. As has been stated in the previous section on the data layer, the data layer belongs to the innermost part of an MES system and thus could well be a poor base on which to install interfaces. Database changes, possibly conversion of the database system and a complete circumvention of the application layer and of the process mapping, are the clear arguments against an interface of this kind.

File-based interfaces

Although somewhat old fashioned from the technical point of view, as far as the applications are concerned file-based interfaces still do the job they should do. File-based interfaces can be set up on the process mapping and will thus in principle be compatible with future releases. Further advantages are the simplicity of data exchange both in the development stage and also in subsequent application, the ease with which contents can be monitored and also the simple possibility of sending processed files to storage for monitoring purposes. Disadvantages are found in the lack of an online capability with file-based interfaces since files need to be retrieved on a cyclic basis. For this reason one preferred use for file-based interfaces is with middling to larger quantities of data whereby a certain amount of time lag is acceptable.

In addition to proprietary file formats (ASCII, CSV, and so on), XML files are being used more and more frequently for data exchange. The advantages of XML files are the platform-independent storage of data and the possibility of exchanging these files within web services.

Web services

Currently everyone in the IT field is talking about web services as a possible cure for a large number of existing inadequacies in software systems. The essential properties of web services are:

- Standardization
- Network or internet-based data exchange
- Platform-independent structure
- Online capability
- Ease of expandability

These properties indicate that web services supply all of the requirements needed for the tasks which system-independent communication has to perform. Due to the standardization implemented by the World-Wide Web consortium W3C, web services offer the necessary long-term stability and goal-oriented development required for an industrial standard. Web services in principle support online inquiries, which means that ad-hoc inquiries in other systems are technically feasible. The way things look today, web services seem like becoming in the medium term the standard technology for data exchange between ERP/PPS and MES systems.

RFC and IDOC

This technology comes from the SAP environment and forms a basis for all online–capable communications interfaces between SAP netweaver and MES systems under SAP. RFC stands for so-called remote function calls – in this case, the means of transportation for the data. For this purpose SAP provides the MES system with a program library as well as a development environment which allows the interface to be created on the MES side. The data are transferred in IDOC format (intermediate document) and the data structure and thus the data contents can be determined from the structure of this IDOC. The disadvantage of this solution is that the program library is made available on the SAP side only for the most important operating systems and that this technology is used only in the SAP environment.

EDI

This interface for the exchange of structured business data is mentioned only for the sake of completeness. EDI (electronic data interchange) has not yet played any major role in communication between ERP/PPS and MES.

5.2.2 Interfaces for horizontal integration

Starting out from the architecture of an MES system, all interfaces use the process mapping to call the defined workflows together with the corresponding data, doing so via the messages which are available (cf. Fig. 5.5). On the basis of this definition an interface implemented in this way offers the following advantages:

- Interfaces thus use the business logic defined
- Even for interfaces there are ways of intervening in the sequences of the workflows on the level of the process mapping
- In this case interfaces offer compatibility with future software releases

With MES systems the user interfaces and data collection systems are normally based on the process mapping. This makes it possible for the MES system to start an operation, for example, at a data collection device and finish it at an information workstation. If the manufacturer of the MES system does not hide the messages in the direction of the process mapping and makes this information available to his customers or partners, new possibilities will then become available on this level regarding how the MES system can be used. The use of these messages is particularly suitable for linking existing systems with a certain intelligence or their own processing capability to an MES system and thereby making a horizontal integration possible. If this idea is followed through to its logical conclusion, the MES system will become an integration platform and existing island solutions will not need to be retired but may instead stay on as specialized data collectors which pass their data on to the MES system.

5.2.3 Interfaces with production facilities

Interfaces with the actual production facilities, such as machine, machine group, or production line, not to mention installation controllers, are indispensable to an MES system.

The most important data which an MES system passes on to the production facilities are set-point inputs, process value inputs, formulations and mixes, and also DNC programs. The most important data which an MES system receives from the production facilities are basically the machine cycle, counting signals, operating signals, machine status, measured values and process data.

The idea behind these connections is, First, to realize a high level of automation and thus to increase economic efficiency and reduce the incidence of wrong operation. On the other hand, requirements concern the attainment and monitoring of a specified quality in the finished product and also the quality and monitoring of the production process itself.

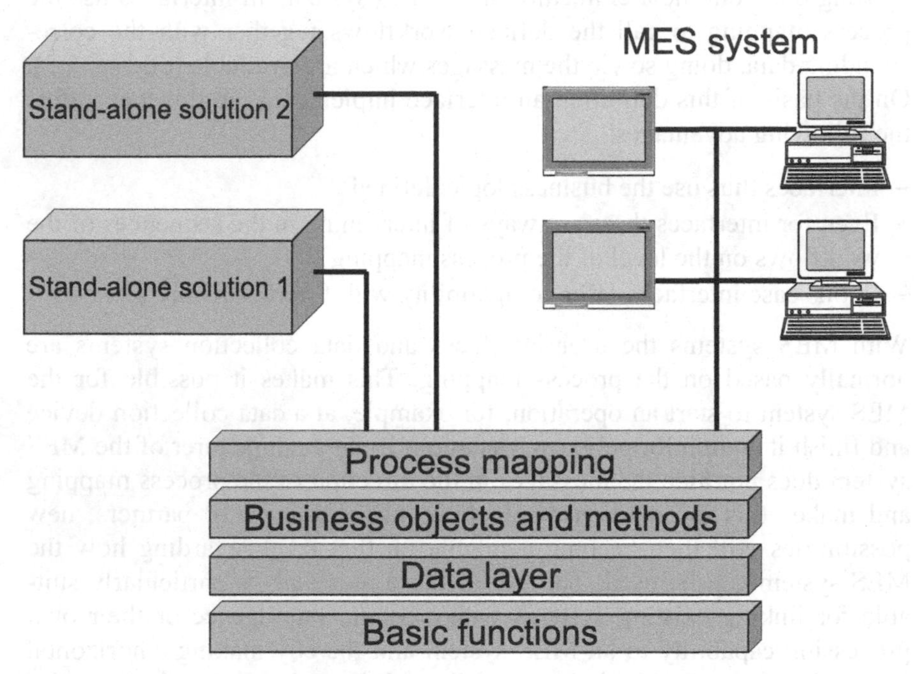

Fig. 5.5. MES as integration platform

Proprietary interfaces

There is a large number of interfaces with production facilities and listing them all here would be beyond the scope of this book. This diversity is however very much a problem for MES manufacturers and customers since due to the lack of relevant standards every machine manufacturer up until a few years ago and in some cases even today is forced to develop his own, proprietary controllers and protocols with the result that connecting up a varied machine park often turned out expensive for the user as well as complicated. In the meantime some MES manufacturers have, as a result of years of project experience, acquired the knowledge required for connecting up the most varied machines and controllers which means that even heterogeneous machine parks can be economically connected to an MES system.

However, since there will be further development even here, in future the machine manufacturers will have to bow to pressure as regards standardized and open interfaces. Not until then will the user have the security of knowing that his investments are secure in the relatively long term and that the costs of connecting a machine into an MES system will turn out as low as possible.

OPC

By OPC (OLE for process control) is meant a Windows-based interface for data exchange between two systems. This interface was specially developed for machine interfacing and has the advantage that, similarly to the case of web services, a consortium (the OPC Foundation) looks after interface definition and compliance. Further advantages of the OPC interface technology include:

- Getting rid of manufacturer-dependency in hardware and software;
- Simplicity in the configuration of the information to be exchanged;
- Network capability;
- OPC permits parallel access to the data provided by the OPC server.

Euromap 63

Euromap 63 is a standardized interface for data exchange with injection-molding machines which was defined by the European Committee of Machine Manufacturers. The Euromap interface has the following properties:

- Euromap 63 specializes in data exchange with injection-molding machines;
- Euromap 63 is not bound to a particular platform. Euromap-capability is rather a property of the controller of the injection-molding machine.

Due to the wide variety of controllers used by the machine manufacturers as also to the fact that the contents of the interface are freely definable by the machine manufacturers, their actual implementation will vary from manufacturer to manufacturer. The degree of abstraction will not therefore be as good as with OPC although it does appear that due to the spread of Euromap 63 the injection-molding field can count its machines amongst those most frequently connected to MES systems.

5.3 User interfaces of an MES system

One fundamental feature of MES systems is their user interfaces. The most important areas of application include, on the one hand, specialized interfaces for use in production. On the other hand, an effective configuration of the system should be possible, as also a clear presentation of the data collected. The section will deal with the technical possibilities and in particular also with the benefits which MES systems can offer users at this point.

5.3.1 Technologies for user interfaces

Local client

This technology exists in several forms, ranging from the simple client/server system to distributed applications in which part of the application is no longer executed in the client.
The advantages of this solution:

- There is good exploitation of local resources;
- Communication only occurs in relation to data exchange;
- Complex applications can be realized without any problem.

The disadvantages of this solution:

- High administrative cost;
- High costs per client.

Thin client

In contrast to the local client, with a thin-client solution only a kind of interpreter is used in the local workplace computer. Both the interface and the data are first sent to the client where they then together define the application. By means of local data storage strategies at the client end, longish waiting times as the data are loaded are reduced.
The advantages of this solution:

- Lower administrative cost;
- Low costs per client.

The disadvantages of this solution:

- Complex applications take a lot of time and effort to set up;
- High requirements as regards the LAN infrastructure;
- Limited possibilities in the graphical representation of technical situations.

Windows Terminal Server and Citrix solutions

These solutions are based on the strategy of combining the advantages of the two solutions just mentioned. Normal applications are used (local clients) which are however run on special servers. The only thing now left running on the local workstation computer is a special client which no longer interprets the data and the application but only receives and displays screen output data from the server and, in the other direction, sends user inputs from, for example, keyboard and mouse to the server. This combination makes it possible for complex applications to be run on less powerful workstation computers. In addition, it delivers reductions in administrative costs since the application only needs to be installed on the server.

Smart clients

Microsoft is currently publicizing its new development framework .NET. With this development framework so-called smart clients can be created. During the development process the developer specifies which parts will run locally and which parts will run on the server, which means that the applications appear to be a mixture of local client and thin client. Another outstanding feature of smart clients is that no separate installation routines are now required: instead, administering these systems is made easier by software distribution by copy command.

5.3.2 User interfaces for configuration, monitoring and reporting

In many cases the operating system Microsoft Windows in its current version of Windows XP is used as a platform for the fields of configuration, monitoring and reporting and is regarded as industrial standard. As regards the technology for the user interface, all of the possibilities we have mentioned are candidates. If a correct selection of technology is to be made, the benefits of the application should stand in the foreground. There is no doubt that web solutions will suffice for information workstations and simple controlling. On the other hand, even in the very near future, a local client will be required by complex applications such as control stations, planning modules or applications with special requirements regarding effective working.

The user interface of an MES should have the following important properties:

- Easy adaptability of the interface for customer-specific wishes by means of, for example, user fields;
- Powerful tools for creating customer-specific reports;
- Powerful tool for creating complete customer-specific applications (application designers);
- As has already been explained previously, here too the correct combination of technology and application offers the best possible benefits for the user.

5.3.3 User interfaces for data collection

Special requirements made of the user interface

In the area of data collection, in addition to the rapid and effective acquisition of data, displaying the actual situation stands in the foreground. As part of the shift of responsibility in production towards the workers (group work, for example) or due to attempts to make savings, reporting in the area of data collection is however becoming more and more important. This means that comparisons with the previous day or week are already possible on the operational level. For the reasons we have mentioned, Windows or Windows CE systems (CE for consumer electronics, systems for smaller applications) is being used more often, even in the area of data collection, since the presence of a graphical interface means that more information can be displayed.

In addition, these operating systems form a good basis for networking in the local network as well as permitting easy inclusion of printers. Good networking capability gives Windows-based systems decisive advantages in the growing segment of WLAN interfacing. In this case the advantages of "mobile data collection" and "online data access" can be combined together.

The data collection devices are operated manually via special keyboards integrated in the housing or via touchscreens. In addition, the most varied reading systems are used for bar codes, RFIDs or other identification vehicles, including even biometric recognition.

One argument used in favor of Windows-based data collection systems is that investment is protected as this equipment is supported by virtually all MES systems. On the other hand, there is a large market for special terminal solutions based on other operating systems which will principally be used when requirements are simpler or when the costs of each data collection device have to be kept low.

Practical possibilities for different technologies in user interfaces

The user interface technologies mentioned some time back find application even in the data collection field. One of the most important requirements in the field of data acquisition is off-line capability: should the server or the network fail, the data collection devices should carry on working as independently as possible depending on the application. This is why in most cases local clients are used (so-called intelligent clients) which control the program flow, recognize a communication failure and buffer the data locally.

Thin clients such as web-based solutions are becoming increasingly important in the field of data collection. Until now the lack of off-line capability in thin clients has greatly limited their use in this field. As WLAN solutions become more popular, a higher failure tolerance and greater mobility is achieved which means that even thin clients can be used in data collection thereby making possible a reduction in administrative outlay.

One current trend is the use of Windows Terminal Server and of Citrix solutions even in the field of data collection. If your network has a high failure tolerance, here too the combined advantages of these technologies can be exploited in production.

5.4 Outlook

Nothing is as constant as change. This saying is more than appropriate even in the field of MES systems. On one side the technologies are changing. As can be deduced from what we have said above, the trend is going clearly in the direction of cost reduction, economic efficiency and saving administrative expenses. The readiness of users to invest in new technologies is no longer oriented by "featuritis" but unmistakably by a different goal: that IT costs must be reduced.

On the other hand, similar demands are made of production in a modern company. Process sequences must be changed quickly, processes must be remapped rapidly and flexibly, maintenance times, downtimes and storage times must be reduced.

Selecting the correct MES system is an important aid in reaching these goals. If the MES system is oriented towards the technical requirements of users, it will allow them to reduce overheads when they use it themselves.

But what is a lot more important is that the right MES system delivers the necessary flexibility to allow the requirements of the user to be satisfied within a clear cost- and schedule-related framework. For this the MES supplier provides an MES system and also trained personnel so it can be adapted as optimally as possible to the needs of the user via

- Parameterization
- Customizing
- Classic adaptation
- The user's own development work
- The functions
- Process mapping
- Interfaces
- User interfaces

We especially mention the simplicity factor with which the process mapping can be adapted since changes in the life cycle of an MES system cause the most costs here. Furthermore, the MES supplier should cover, both technically and on the application side, as many as possible of the real requirements of the user. If this is the case then nothing should now stand in the way of a partner-like relationship between the right MES supplier and the user, especially as these are likely to be medium- to long-term relationships.

6 Integrated production management with MES

6.1 MES systems make production management possible

While workplaces in the office have been equipped with PCs for the last twenty years we are now experiencing the systematic continuation of this trend in the field of production. The production worker's office desk comes into existence at the information point in production.

In contrast to the way computers are used in management where the emphasis is on word processing, spreadsheets and ERP systems, the production worker uses programs tailored to production-related data handling.

The MES system is the information and data collection system for manufacturing and production. The very latest status information about orders, machines, tools, materials and personnel is brought together here and prepared appropriately for the relevant applications. MRP controllers, production schedulers, logisticians, production foremen and the workers themselves use the online information from the MES system and can make the right decisions for the particular situation when production processes do not run as planned. Optimum production management was not possible until the arrival of MES systems.

The ERP system looks at production through a telescope as it were. It plans and recalculates production sequences on the basis of production status messages from the MES system.

Process computers displaying visualizations are island solutions optimally tailored to the process in question. From the point of view of production management this is like looking through a microscope. The MES system is the connecting link between the ERP level and the production process. All of the data which allow production management to make the best decisions are brought together here.

6.2 The MES model

MES systems compare acquired data against target requirements. This means that MES functionality divides into a data acquisition functionality and a functionality for monitoring these acquired data. The MES system is deployed in a targeted manner and focuses on the data which are important for the company in question. The present treatment can be divided into the three central topics:

MES data analysis
The task of data analysis stands at the beginning of an MES project. In many cases the typical MES data entry objects can be derived from the user's production process or the sector of industry. The result is a listing of the data entry objects to serve as the basis for the database of the MES system.

Data acquisition functionality
The second area of focus in MES requirements are those relating to an optimally tuned data acquisition functionality. What degree of automation can be implemented in data acquisition? Which data will need to be entered manually? At this point it is decided which data acquisition functions can best be integrated into the existing production sequences.

Information display and preparation of analyses
MES users are interested in different MES data entry objects at different times in different cumulations and differently aggregated presentations. The MES system must offer information editing functions to every user and pass collected information onwards via interfaces.

The relationships between data acquisition functionality and the analyses are shown in the MES model. Data are registered at the data collection terminal by means of the relevant acquisition functions for the data entry objects in question and taking plausibility checks into consideration.

The MES system administers the data in a database which serves as a data archive for the overviews and analyses.

The data acquisition functionality works in dependence on planning inputs relating to the corresponding reference variables – for example, requirements for order and article. The personnel steer the data acquisition dialogs, making use of selection lists or making manual inputs. Context-sensitive plausibility checks are made here against the MES database.

In some cases the process interfaces with machine or weighing equipment controllers are triggered by manual inputs. Current and objective data are taken directly from the process and evaluated in order to obtain a description of production progress.

The decisive advantage which MES has over traditional data collection systems such as written records (which are still used even now) is that the time dimension in value generation can for the first time also be considered and registered. In the light of the fact that an overwhelming majority of resources represent potentials which are used in a manner proportional

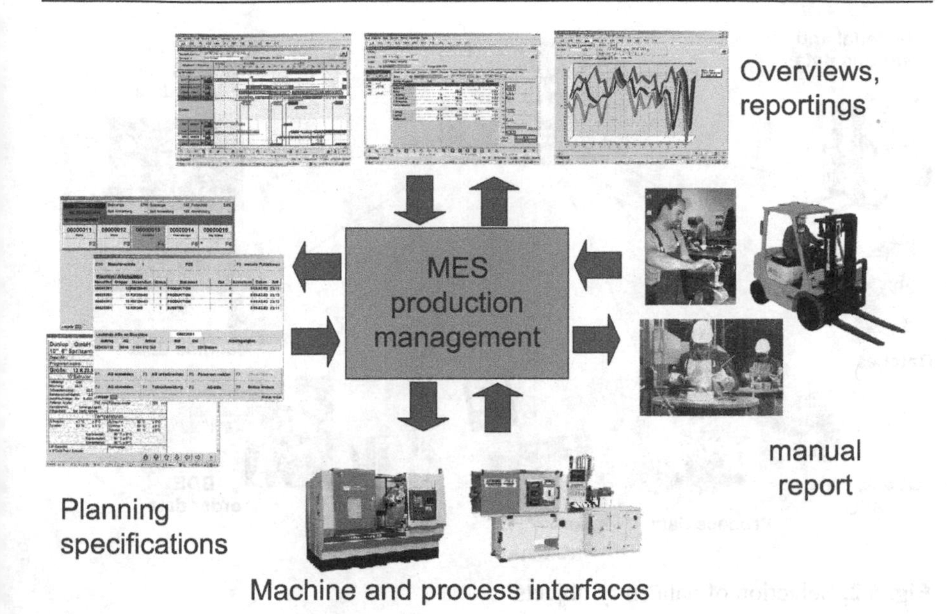

Fig. 6.1. The MES model

to time, a mapped image of these dependencies becomes an important basis for decision-making.

6.3 Data analysis: information in an MES system

The modern data acquisition terminal is integrated into the corporate network (intranet) and enjoys full access to those company datapools relevant to production. The data acquisition client is therefore a powerful tool and, since it communicates with the peripheral device which has been tailored to its environment, is also a regular "factotum". The central task of the MES is to focus ergonomically on the major data objects and to handle the data ergonomically.

Anyone planning to introduce an MES system must first be clear about which data objects he is going to set his data acquisition up on.

The relevant data model used for the MES system will be guided by the sector of industry and the production processes. Depending on the type of use, other data entry objects will be of interest: the *plant and equipment manufacturer* will be interested in time data and hardly at all in quantities while the *repetitive manufacturer* regards process speed at the machine as the essential factor. Quality inspections, on the other hand, will have equal importance for both of these target groups.

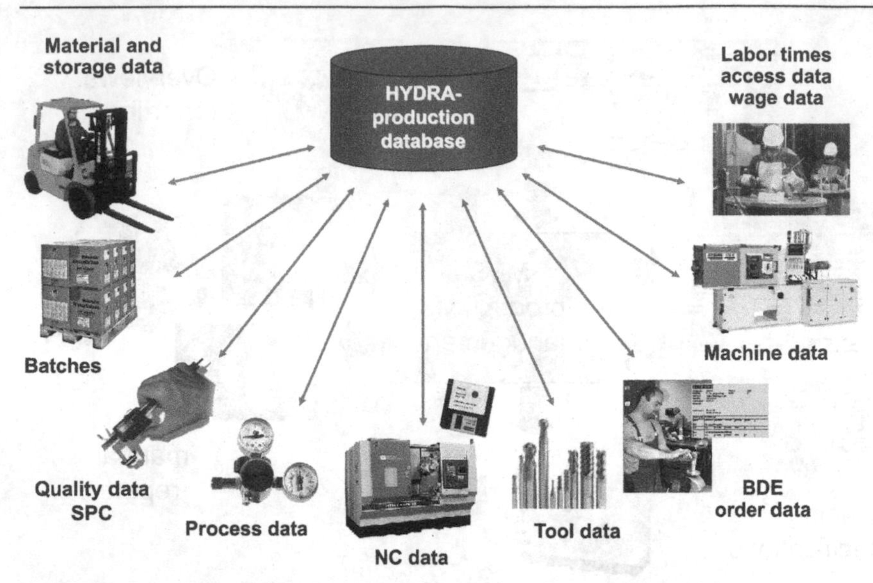

Fig. 6.2. Selection of data entry objects

From the ERP system the MES system receives input requirements regarding the order and the resources scheduled for the production process. These data are administered in corresponding inventory charts. The MES system does however also have MES-specific configuration data which are administered locally in the MES system. The MES system derives the requirements applicable to the production processes from order inputs and from the configurations.

Status signals from the production process are assigned to the MES data entry objects. The data are compared with the target requirements and support production management. The MES system passes cumulative results on to the ERP system.

Decisions as to how long data will remain in the MES system are made for each data entry object as also regarding incorporation in the MES data model, the origin of the master data, use and properties within the MES system, and feedback to the ERP world.

6.4 Operating resources: machine or installation section

The central reference point of classic production data acquisition is the machine, the machine group or the plant section. Defining the individual work centers is process-dependent in the MES system and will therefore only be oriented to a limited extent by the requirements of the ERP system. With the MES system being based on a more finely implemented machine definition, a production line in MES will therefore in most cases be broken down into individual machine groups since the process steps active there form the basis for data acquisition.

The production speed, the quantities produced of good or poor quality, and also the reasons for and durations of idle times are the main data which are acquired at the machine.

6.4.1 Order/operation

An order originates in the ERP world and with its plan requirements determines the targets for the production process. At the same time it serves as a cost collector for recording services performed (times) and quantities produced. For this reason the order represents the backbone of data acquisition and is thus the classic data entry object.

Anyone in the field of data acquisition talking about the order in most cases means the operation, the operating sequence or the activity. The operation includes all information which according to the work plan relates to the corresponding work center and to the process step within the order.

The work operation transports production information to the data acquisition station. This data includes master data from the work plan (standard speed, planned work center, planned set-up time, planned running time, work instructions) and from the material master (bill of materials information, required quantities, drawings) as well as the descriptive data for the order (such as target date and target quantity, and possibly customer information, printed text for labels, and so on). The work operation also determines resource requirements, which in production are assigned and plausibility checked either implicitly or by manual inputs as an MES signaling object.

At the same time the work operation receives information from the production process about the progress of production and displays this at the data acquisition station, transfers information into the MES database and collects feedback data for the ERP system.

In addition to pure production orders, an MES also processes reworking orders, project orders, overhead cost orders and inspection orders, which may very well be handled differently within the data acquisition process.

6.4.2 Material

Data acquisition may relate not only to those materials which flow into the production process but also to the material produced. In the case of incoming materials, in most cases plausibility checks are carried out against the bill of material. If a manufacturer is required to register and track on a batch basis, he will need to identify the material batches in the production process at the data acquisition client. Discrete quantitative consumption which is measured at the work center also represents a material-related report which can be passed on to materials management.

Manufactured materials are logged as yield, scrap and reworking quantities. For concatenated processes with charge-based data acquisition the MES system generates unique batch or lot numbers for further tracking purposes. Materials can be barred directly in production and in this case cannot be used further until a decision is made about their future.

6.4.3 Resources and production tools

Resources which are required for carrying out the production process and which are simultaneously only available in limited capacities are included in the planning, are assigned and then taken up by the production process. This includes the tool, specially trained personnel, such as the line engineer or the quality representative, as also special handling devices or required items of equipment.

In many production environments the tool as a resource is even more important than the machine. Tool-related maintenance is based on registering service times and quantities (work cycles). Recording the tool number will be an unavoidable requirement when several tools of the same tool type are being used, such as is normal in large quantity production, for example. On the other hand it will not be necessary if only the planned tool is used.

Active resources are barred with respect to parallel scheduling and use in other machines.

Machine programs or machine settings records are special resources which the terminal transfers into the machine controller during the set-up procedure. Monitoring release criteria and versions management are also tasks the MES system must perform for these resources.

6.4.4 Process values

In the case of highly automated processes and in production environments in which product quality is markedly dependent on individual process values, a central role is played by the recording and the permanent monitoring of characteristic process values. Here MES data acquisition records the relevant signals direct from the process (analog or digital), displays them and saves them in the database, either on the basis of specified random sampling patterns or at defined intervals. In the case of batch-based production there is frequently a need to save recorded process values in relation to the batches of material produced.

6.4.5 Personnel

The aim of data acquisition relating to operating personnel is to be able to assign the work done to a cost unit such as, for example, the production order, the maintenance or overhead costs order. Personnel-related messages in production are registered in an integrated MES system combined with clocking-in and clocking-out data. This means that attendance times can be compared with productive working times at the workplace. If personnel are working on the basis of performance-related remuneration models, the worker's cost-center–related or order-related messages will form the basis for calculating the corresponding time balances for the incentive payment or piecework models held in the system.

Target/actual comparisons, such as current personnel overviews, for example, are based on the personnel messages of the workers.

Identification of the individual plays a central role in the implementation of access control requirements. In this regard the MES system supports not only conventional card readers but also special data acquisition techniques for checking biometric data, such as fingerprint readers or even retina scanners. The job of the MES system is to manage the reference bit pattern for each person for comparison purposes during the identification process.

6.4.6 Inspection and testing characteristic

The inspection order tells a production process at what time or quantity-related intervals individual characteristics of the material being used or manufactured are to be subjected to inspection. The job of MES data acquisition is to record the inspection results, to compare target values and tolerances, and also to display characteristic curves, for example, in the form of control charts, Pareto analyses or analyses from statistical process control. What is particularly important here is one particular function of some MES systems whereby irregularities or deviations from target requirements are immediately displayed upon data input or measurement.

6.5 MES data acquisition functionality

From the DP point of view, the data acquisition functionality of an MES system is an interface which has been shifted into the production environment. In most cases little attention is given to the importance of this interface although a well-functioning data acquisition interface is of decisive importance to the acceptance of the MES system.

The great variety of data entry objects and their automated identification, the ergonomics of dialog control, plausibility checking of acquired data, techniques of data acquisition are some of a wide range of aspects which have to be taken into account when planning and implementing an MES system if a reliable interface is to be obtained and the MES system optimally fulfill its objectives.

One requirement which every company makes of data acquisition is that data should be acquired in production without practically any additional effort on the part of the worker. As utopian as this demand basically is, it must nevertheless be consistently taken into account in equipping data acquisition stations and in the design of data acquisition functions.

Data acquisition is heavily affected by a number of system requirements which ensure good data quality:

- **Ergonomics.** Efficient operator control on the part of the worker is a necessary condition of success in implementing an MES system.
- **Plausibility and completeness of the data.** Constant checking of consistency results in a high level of process reliability. Plausibility and completeness guarantee a high quality of data and are thus of decisive importance to the benefits of data acquisition. In addition, the least possible reworking due to correction or cancellation of the acquired data is ensured.
- **Operational reliability.** The off-line capability of the data acquisition program and the possibilities for buffering acquired data ensure the high level of availability of an MES system.

The job of the data acquisition terminal is the provision of data acquisition dialogs for the worker for the purpose of controlling production sequences, data transportation of planning requirements for the data entry object concerned (order, tool, material, personnel, and so on) in production, and recording process parameters via the corresponding interfaces with the process. Other tasks include the ergonomic presentation of information at the terminal and also the operation of system peripherals, such as, for example, printing accompanying documents or labels. Further system functions support the assignment of unique identifications, for example, for identifying material batches.

6.5.1 Data acquisition terminal equipment

Ergonomics and security is achieved by the use of identity readers and by data acquisition which uses as little paper as possible and is as automated as much as possible. Local conditions in production, distances from the reporting station, temperature fluctuations or a dirty environment are factors which determine what is the best equipment configuration at the data acquisition stations.

A wide variety of technical equipment caters for the range of requirements of the most varied industries and the most varied manufacturing processes. Ergonomic data acquisition by the MES system is supported by terminals with touch screen operation, mobile data acquisition devices with a wireless-LAN connection, electronic readers and scanners, scales and machine or installation controllers which condense the data and provide the corresponding interfaces.

The type of data acquisition must be specified for each data entry object. For example, an order can be selected from the work-center–related planning list, signaled via a bar code on the production document or assigned manually. The MES must make all possibilities available and support an individual configuration for each work center.

How the data acquisition terminals are equipped and which peripheral devices are selected must be guided by the means of identification used for the individual data entry objects.

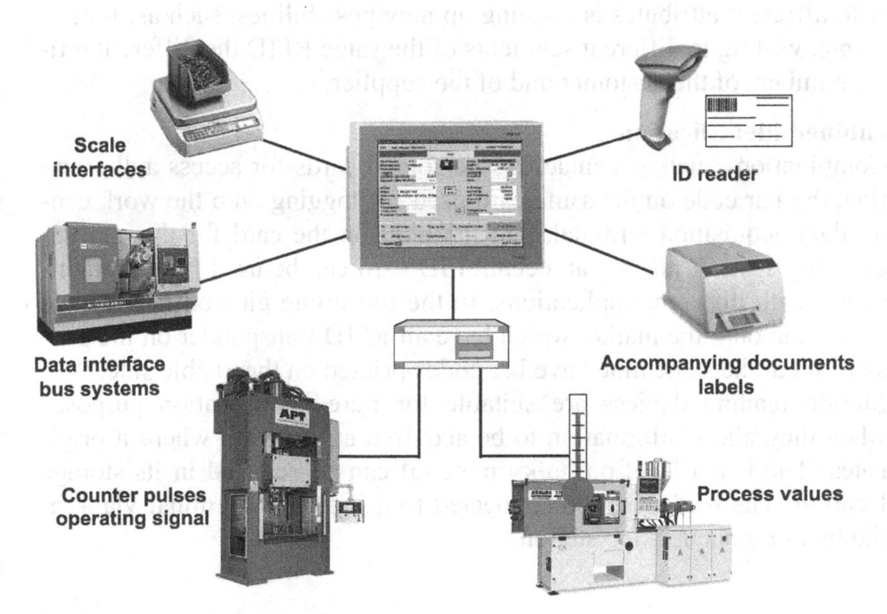

Fig. 6.3. Definition of the data acquisition infrastructure

Stationary PC-based terminals with touch screen operation.
PC-based terminals make all possibilities available to the data acquisition application. Large displays and touch screen operation make ergonomic operator control possible and are ideally suited to presenting information in written or graphical form. Selection lists can be clearly presented in the data acquisition dialogs. Peripheral devices with corresponding driver programs are universally available. Wireless-LAN connections mean that PC-based terminals can be used flexibly.

Mobile terminals.
Wireless-LAN equipment permits deployment of mobile terminals which combine local flexibility with online plausibility checks against the current database. Selection of the terminal hardware and of the corresponding operating system will need to be matched to the application in question. Alongside PC-based terminal architectures, PDAs or even cell phones are playing a greater role in mobile data acquisition. In most cases mobile terminals are combined with the corresponding reader devices.

Contactless identification.
Under the buzzword of RFID (radio frequency identification), contactless cards for individuals, transponders for identification of materials in the most varied shapes and designs, suitable for even the harshest production environments, are delivering a new flexibility in the identification of data entry objects. The flexible description of different segments with different attributes is opening up new possibilities, such as, for example, writing to different segments of the same RFID the different article numbers of the customer and of the supplier.

Combined identification.
Combinations such as contactless personnel cards for access authorization, the bar code on the same card used for logging onto the work center data acquisition terminal, and the chip on the card for the canteen charging system, mean that a central ID card can be used for communicating with different applications. In the meantime glue-on labels have also come onto the market which have an RFID transponder on the glue side and at the same time have bar codes printed on the visible side.
Mobile reading devices are suitable for pure identification purposes when they allow information to be acquired at the place where it originates. The batch label on bulky material can be scanned in its storage location. The reader can be connected to a stationary terminal via a radio link or via a docking station.

Biometric data.
Identification of individuals in sensitive environments or as part of access control can be effected by recording and comparing biometric data. In the meantime fingerprint readers and even retina scanners have become available at a reasonable price thereby rendering a wide use of these identification techniques no longer utopian. Particularly in the field of access control it is necessary to be able to control door openers. In this case the terminal supports the corresponding interfaces.

Printers and printed codes.
Bar codes and even matrix codes are easy to create and flexibly configurable. For this reason printing labels at the machine's data acquisition station has further gained in importance. Its use in production logistics, for the identification of material batches or transportation containers during reclassification or relocation procedures, has become necessary in many sectors of industry due to statutory requirements and is gaining more ground.

Alarm devices.
There is a necessity for an alarm facility in the production process directly at the data acquisition terminal. The events triggering an alarm are not only displayed by the MES system but also take the form of acoustic or visual alarms. Configurable alarm situations are communicated by the MES system directly to the responsible persons by email, pager or cell phone on the basis of specifiable escalation levels.

Web clients for data acquisition at remote locations.
Under the buzzword of the "extended workbench" more and more production environments are working with decentralized production facilities or external service providers. Transparency for production control throughout the entire production process will not be possible unless the external production processes are also included in data acquisition relating to production progress. In addition to being fitted out with identical data acquisition equipment, modern MES systems offer the possibility of data acquisition via a web client. Configurable data acquisition dialogs which can be accessed on virtually any PC with an internet connection permit data to be collected on an inter-company basis.

6.5.2 Information for the worker

As automated processes in production become more important the worker needs more and more information. The right information at the right place, comprehensive, fast and presented ergonomically – these are the requirements which the MES system must meet when supplying information in the production environment.

The data acquisition terminal displays the current information regarding the status of all data entry objects, thereby creating transparency in the workplace and ensuring process reliability. Examples of how this need for information manifests itself are current status monitors for the state of the machine and the production process, for the order, the operation and the material batches being used, for the machine operator or the team, for pending inspections and recent inspection results, for the machine program currently loaded, for tool use and upcoming maintenance intervals for machine or production tools.

Since the data acquisition terminals are networked, there is practically no limit to how master information can be presented: images of assembly drawings or finished articles, work instructions in the form of video sequences as an assembly aid, the display of setting parameters from the DNC programs by means of suitable viewers are only some examples of the plethora of information which is made available to the worker. What is of decisive importance is the up-to-dateness of the information presented and, unlike printed documents, this can be guaranteed at all times. Evaluation functions for the last shift, the utilization ratio of the previous week,

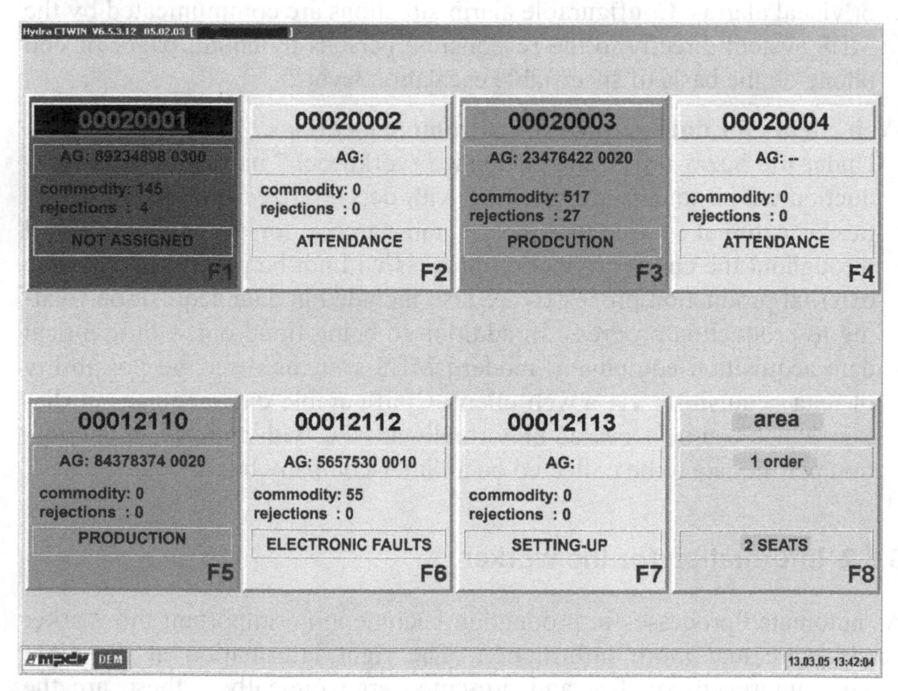

Fig. 6.4. Presentation of current information at the data acquisition terminal

consumption of materials during the current month, and so on, round off the range of functions offered by the data acquisition terminal.

The personal motivation of the worker plays a decisive role in the optimization of production processes. Information about personal needs and personal targets are major factors. This includes the current time balance of the employee, the current level of his incentive wages as well as his success in reaching targets on the basis of the individual employee target definition. These displays of information aimed at the individual are becoming more important, particularly with the arrival of manufacturing scorecard methods (Kletti J, Brauckmann O, (2004)).

The MES system architecture provides tools which are used for the preparation of information and for designing layouts for how the information is to be presented. This means that all information functions can be individually designed and via a log-in function can branch off into a personalized information pool for the employee.

6.5.3 Modularity supports the diversity of data acquisition dialogs

Acceptance by employees is of decisive importance to success when an MES is introduced. Most of the employees in a company will be working with the data acquisition terminal. For this reason, decisive potentials with regard to system ergonomics may be found in its design and dialog control.

Modern MES system architectures support a flexible adaptability of the data acquisition functions to the data acquisition process. To make it possible for the standard system to be customized to the individual process, the system architecture itself must provide the necessary capabilities.

It must provide modularity and adaptability of the data acquisition dialogs to the individual work center. Very, very few companies have a homogeneous production structure and the homogeneous data acquisition structure this implies. Injection molders together with their tool making departments thus bring to an MES system the requirements not only of a repetitive manufacturer but also those of the one piece make-to-order manufacturer. With such a production environment it must be possible to design data acquisition dialogs individually for each work center.

Configurability with regard to the operator (for example, an option allowing foreign workers to select a language) represents one more demand on the flexibility of dialog design.

The wide variety of configuration settings in an MES system, the possibilities for customizing the data acquisition functions, and the individual design of dialogs for each work center, all of these require a modular system architecture.

Integration of the data acquisition interfaces into the dialogs must be configurable by, for example, specifying that the "personnel number" field can only be occupied by the LEGIC identification number or that a quantity status report field can only be filled out from the weighing device controller.

6.5.4 Plausibility in the data acquisition process

The MES system will secure a good quality of data when a plausibility check can be carried out on the data directly during the data acquisition process. Depending on the data entry object in question, the most varied tests can be defined in the MES system. Static checks will be concerned with the master data or set-point data (for example "the actual quantity must not exceed the set-point quantity") while dynamic checks relate to the course of the production process (for example "with overlapping production the quantity produced in an operation must not exceed that of the previous operation").

With manual dialog control, infringements of plausibility, such as exceeding quantity levels, can be displayed to the production personnel directly. This allows the worker to react to the infringement by correcting his mistake and inputting the correct quantity. In this way the need for off-line corrections can be largely prevented.

On the other hand, plausibility checks in communications with machine controllers are a problem which can scarcely be solved in the dialog. The interface between process and data acquisition client must deal with this violation by means of a suitable exceptions handling routine and without impairing data quality or integrity.

In many cases checks only make sense in certain parts of a company while in others they are utterly useless. For this reason modern MES systems support dialog-oriented plausibility checks which by means of customized settings can be activated or deactivated for different production areas (cost centers, work centers, machine groups) or for different types of order.

One important requirement is checking the status of all data entry objects by means of plausibility checks on the part of the MES system, thereby securing an integral process reliability. For example, should a lab check on material give a negative result, the material in question will be immediately marked as barred. This online status check in the data acquisition dialog before any use is made of materials ensures that material just rejected in the MES system can no longer be reported as available for production.

6.5.5 Which interfaces with the process are best used?

On the data acquisition level modern MES systems support interfacing possibilities which can be used universally. In many cases the machine park to be interfaced with consists of a mixture of old and new machines whose controllers provide different methods of communication. The modular nature and the configuration options offered by the MES play a major role in process interfacing in particular.

Clock signal and operational signal acquisition
Simple acquisition of clock or operational signals via digital I/O or counter inputs still represents an important basis for MDC data acquisition in automated production processes. The large amount of older production machines will in particular continue to determine even the future use of this universally applicable and effective method of machine interfacing.

Machine control protocols
A more complicated method is to communicate with the machine controller directly. This enables the data acquisition terminal on the one hand to transfer data directly into the controller (such as a machine program for machining a particular article) and on the other hand to receive data from the controller about production progress or about machine idle times which have occurred, or to receive any process values desired. In communication the terminal must respond to the controller version and the protocol being used.

Weighing devices
Weighing device interfaces are important aids in automated data acquisition. In most cases the impetus for recording a weighing result comes from a user dialog. Data transfer from the weighing device controller to the data acquisition terminal – a potential source of error with human intervention – is automated and protected. The same general constraints apply to communication with the weighing device controller as to the machine controller interface.

Process communication by OPC
Data transfer by OPC (object linking and embedding for process control) is used for direct communication with the machine or installation controller. The OPC technology provides a standard protocol which supports data transfer between the controller and external communica-

tion partners irrespective of the equipment manufacturers. MDE signals from the machine, such as counters, clock cycles, status or process values, are just as feasible as data transfer into the controller. This type of communication is not possible without support from an OPC server via the machine or installation controller. In this method the data acquisition program communicates via an OPC client with a specified address space in the machine controller. This means that the data acquisition program can on the one hand access via defined OPC variables all information held in this address space and on the other hand describe data fields in this address space. The task of the OPC client is to make a logical connection between the application programs and the OPC variables.
One advantage of OPC communication is that it is independent of protocol and controller versions. It does, however, have the disadvantage that communication via the LAN, which is used frequently, calls for high network availability.

Special protocols
In some branches of industry special protocols have gained acceptance whose inclusion the MES must support, provided this has already been made possible on the machine side. One example of this kind of standardization is Euromap protocol E63 for injection-molding machines from all leading machine manufacturers.

6.5.6 Data correction in the MES system

All data which are acquired by the MES system must also be modifiable in the MES system. This requirement relates not only to data which were manually input but also to data obtained from process interfaces.

This modification functionality must be backed up by the corresponding authorization checks and when the function is used this must be logged in the MES system.

In the event of a correction the MES system must take into account all interdependencies between the data entry objects. For example, correction of a quantity entry will affect all relevant data entry objects, for example, machine, order and personnel. When corrections are made, all interfaces with higher-level systems must be supplied with the corresponding cancellation records.

6.5.7 Availability and reliability of the MES system

An MES system secures the system availability of the information system for production and thus offers an additional possibility of securing against failure of the ERP system. Practical experience has shown that over a period of several hours of failures on the ERP level, the store of orders held in the MES system enabled production to keep going and this meant that it was possible to avoid major economic losses.

The MES system itself has safety mechanisms which can bridge over a failure of the corporate network. The data acquisition terminal caches data when there is no connection to the MES server. In the event of the LAN going down, plausibility checks at the data entry object against the locally available data pools are possible.

Even with redundancy built into the system, power outages, database failure or system hardware faults can result in total failure of MES data acquisition. Should this occur, the organization goes into emergency mode. For this eventuality the MES system provides a functionality whereby data can be efficiently input later on.

6.6 MES information for production management

Good and objective decisions can be made without having reliable information at your disposal. With the MES system the very latest information about the manufacturing environment is guaranteed available at all times at the work centers of MES users. This information includes displays of the online status of the order, the machine, the process, the tool, the material currently being used and produced, as well as the quality and the personnel.

The MES system provides each target group with objective information based on different time scales. For example, for a particular order or article the MES receives different information under different time aspects: planning lists for the next shift, operations in progress at the work centers of the machine park, orders completed the previous day, or displays of actual times required in comparison with the specified times for all orders for a special article in the previous three months.

The worker at the machine is however interested in different information being displayed than the production manager is. For this reason all data acquired can be cumulated on the basis of different time bases or by departments and graphically prepared. It is important to the machine operator that all machine states recorded are available for individual analysis for group meetings in the same way as utilization analyses serve as a basis for the department supervisor's reporting to the production manager, and the plant OEE figures (OEE = overall equipment efficiency) enable management at corporate headquarters make an objective comparison of different plants.

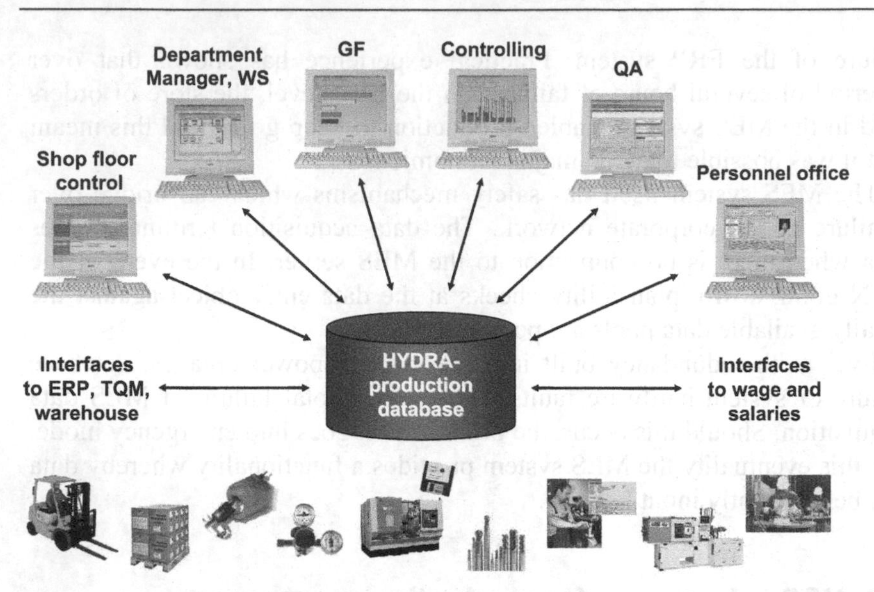

Fig. 6.5. Information displays for MES users

6.6.1 Transparency due to MES actuality

The MES system creates transparency by displaying online information about the current situation in production. Permanent status information and the target/actual comparisons based on this for *all* data entry objects:

Current status overviews of the order, operation, machine and machine group, tool and tool component, production personnel, consumables and feedstock batches, machine programs being used, dynamic process data and current inspection results.

The use of this online information yields potential benefits in production planning and control. By providing a comprehensive pool of information the MES system helps the production scheduler make the right decision when rescheduling is necessary by displaying directly the corresponding effects on competing requirements.

Permanent status adjustments in the MES system result in updated displays for all data entry objects:

- Order overviews with current target/actual comparisons
- Machine status overviews showing the current status
- Changes in material stocks
- Personnel status overviews and updated short-term manpower planning
- Updated tooling lists
- Tool requirement lists
- Updated lists of maintenance activities pending for each resource

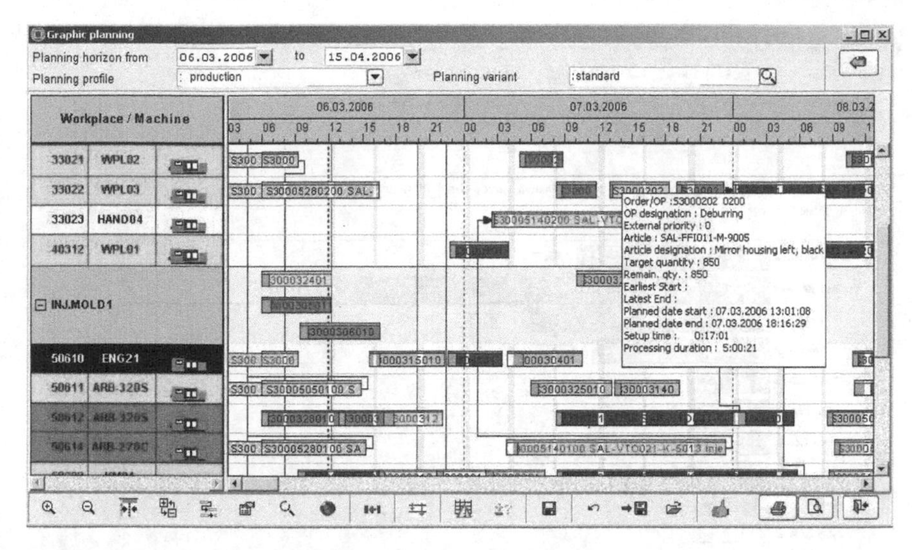

Fig. 6.6. Current machine status and scheduled operations

Alarm situations arising from critical changes in production can be displayed by the alarm management function of the MES system and sent by email or cell phone directly to the persons responsible. By means of escalation levels a workflow is generated in the MES system which results in critical situations being dealt with in an orderly manner.

6.6.2 User-focused analyses

In the MES system analyses can be produced for all data entry objects and over any period of time. Analyses are made available at all data entry objects in the MES system. The following examples will provide just a small sample:

- Article, order and operation profiles
- Analyses of transportation, queuing and storage times within production
- Display of overhead cost balances from registration of all secondary processing times
- Idle-time analyses for machines and installations over any time period
- Comparisons of productivity indices
- Personnel information relating to completed activities
- Incentive wage curves for individuals or incentive payment groups
- Actual-value graphs for process values
- Traceability analyses for material and batch tracking in production
- SPC analyses of defined quality characteristics and scrap analyses
- Tool histories showing service lives

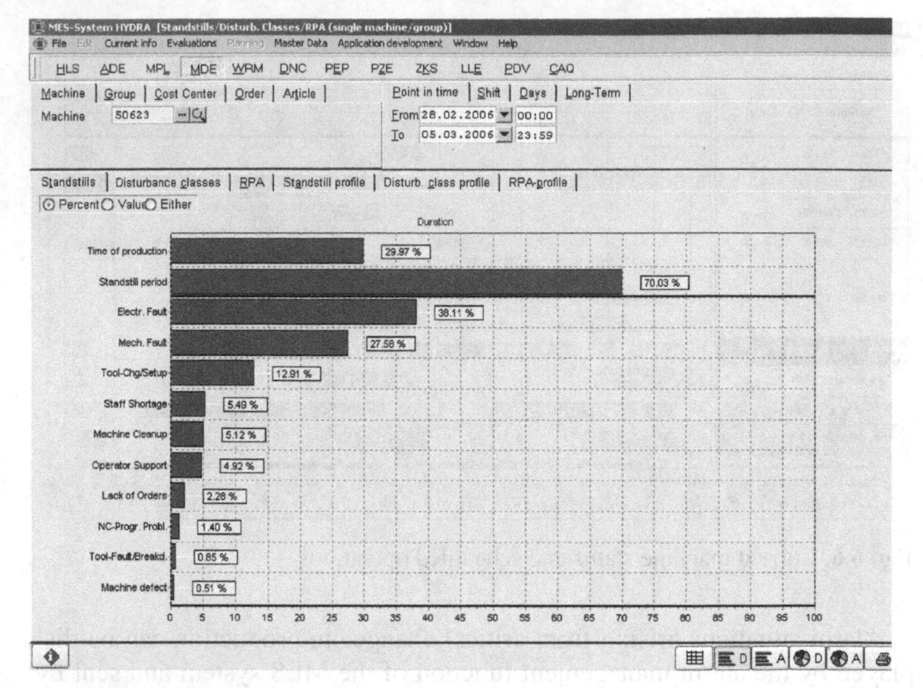

Fig. 6.7. Machine time analysis

Powerful MES systems will even supply the available analyses to a web client. The MES system can thus be used globally as a tool: the production manager can also view at an outside location all relevant information for his field of responsibility and he can directly compare key data across different companies. Manufacturing companies whose structure is changing due to globalization and relocation of production facilities will profit from being able to make inter-plant trend comparisons on the basis of data from the MES systems in those individual locations.

After a configurable time interval of several months the MES system will normally delete all transaction data from the active datasets. Acquired information to be held in long-term data storage in the MES system is combined and transferred into archive tables. Analyses of these archive data can also be included in current analyses by means of the corresponding switches.

6.6.3 Production-related target definition

In its analyses the MES system does not present static data but rather shows temporal changes and current values in the form of a trend curve. For specified operational target agreements the MES system displays

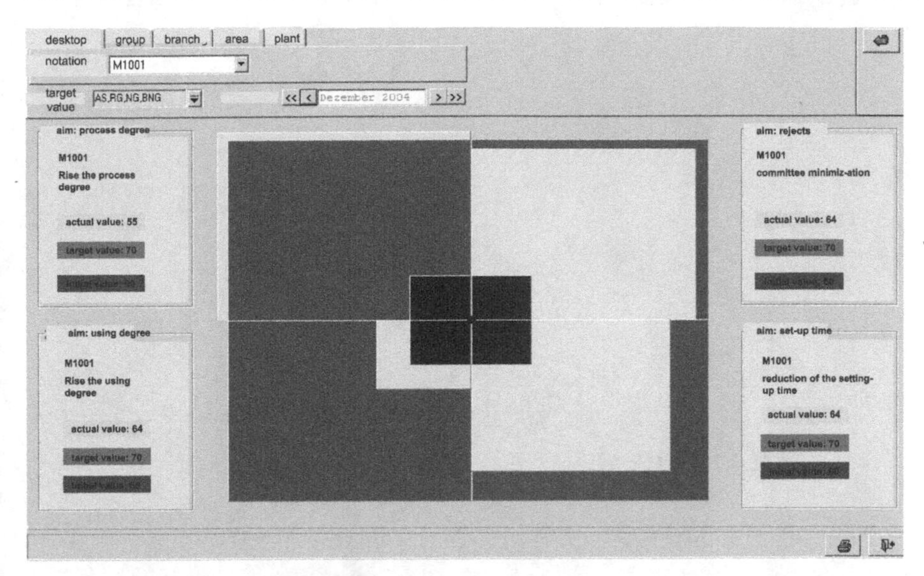

Fig. 6.8. The manufacturing scorecard of the HYDRA MES system

intermediate results occurring along the road to the target and thus the degree to which the target has so far been attained.

The MES system can be used for formulating operational target agreements. The data entry objects and logical linkages between them form the basis for this. Examples include the specification of target values for machine idle-time minimization, for scrap rates, or the formulation of utilization rates which must be reached within a specified time.

Operational target agreements can be broken down by any type of organizational unit. Requirements may be laid down for the plant, the cost center or the work center, the incentive payments group or the individual employees.

The target is published and the actual values on the road to hitting the target are systematically tracked. Everyone involved in this innovation process can read off from the MES information displays how far targets are currently being met. Personal target displays and their attainment rates are integrated into the employee information display.

The MES system provides sustained support for the organization since long-term employee motivation is guaranteed by the target values specified.

Literature

Kletti J, Brauckmann O (2004) Manufacturing Scorecard. Gabler, Wiesbaden

7 Detailed planning and control with MES

7.1 Overview and goals

7.1.1 Incorporation of detailed planning and control

In a manufacturing company the actual production planning is integrated in the general framework imposed by the industry in question and by corporate policy. Commencing with the strategic alignment of a company and the sales planning deriving from this we obtain the general framework for production planning. Investment and growth planning will also be part of this strategic planning in order to provide the necessary production capacities for the production processes.

This brief introduction should make clear that demands relating to the temporal validity and precision of a plan and also the planning problems bound up with it will differ very greatly depending on the level (rough or detailed planning) and the point of view (operations planning and scheduling/production control) of the target group in question.

As regards production planning, here too the goal- and need-oriented approach of a multi-level production planning system has proved itself to be a good basis for an integrated production management.

To enable a better understanding of the terminology and of the detailed planning in MES systems (this will be examined in detail below) an overview is provided of the different planning levels:

1. Tactical planning (sales planning)
Definition of general constraints such as planned investments or even growth targets.

Specification of basic data such as the shop calendar and of production targets derived from this such as, for example, minimization of warehouse stock.

Time frame → long-range, strategic planning
Validity → annual; if necessary, quarterly review
WHO → management/top management

2. Rough production planning
→ Long-range/medium-range
Implementation of strategic planning in primary requirements taking concrete requirements into account (purchase orders/release orders). As part of material requirements planning, ways of meeting incoming primary requirements are determined as are the secondary requirements deriving from the bill-of-material explosion.

In the case of in-house production, the result is concrete production orders and thus the requirements to be planned into the production planning. Depending on the vertical integration or the complexity of the operation sheets the individual production orders will in some cases be linked together into order networks in order to map production-relevant dependencies.

A rough schedule will be drawn up in the ERP/PPS system which takes into account the earliest starting date and the latest finishing date (delivery date less a safety margin).

Sometimes this rough scheduling involves so-called capacity leveling in which a rough comparative check is made as to the feasibility of the resulting production orders. For this check the performance of production, for example, is expressed in quantities of material per unit of time and in many cases only the final fabrication stages are taken into consideration here.

At this point it must be clearly pointed out that this check can only be made against an assumed, abstract capacity and without taking technical circumstances into account. This means its informative content should be assessed correspondingly.

This much too approximate estimate brings with it the risk not only of setting unrealistic target dates or even making unrealistic promises to the customers but also of the orders on hand being planned unrealistically for production. The orders on hand can be arranged more realistically for production with the aid of correspondingly thought-through concepts for optimizing rough planning in a company.

Unfortunately what is still seen in practice, even now, is that in many ERP/PPS implementations only lead-time scheduling is carried out. This is done on the basis of infinite capacity, which means that the feasibility decision is made in the heads of those familiar with the process or even left for production to make.

3. Detailed planning
From the roughly scheduled production orders come the corresponding capacity requirements for the resources needed. Here a distinction is drawn between the primary resources (machines/installations) and the secondary resources (personnel, tools, and so on).

As part of detailed planning, the roughly scheduled production orders, which are competing with each other, are assigned to the capacities available. This type of loading means that realistic statements can be made regarding target dates and the loading situation.

Orders which have been firmed up as part of detailed planning are then passed on to production for execution. Status reports from production are fed back into the detailed planning, thereby allowing permanent comparison of the plan with what is really happening. Since the reality will deviate from the plan with a probability approaching certainty, detailed planning must respond to deviations of this kind.

7.1.2 Tasks of detailed planning and control

As part of detailed planning, roughly scheduled orders or operations will be assigned to the resources currently available, this being done on the departmental or group level – depending on the organization.

Unlike rough planning on the ERP/PPS level, production, which is the implementing body, finds itself faced with the current reality. This reality manifests itself for example in:

- Technical problems in the production facilities (machine idle time, tools breaking, and so on)
- Problems in processing the material being used
- New priorities in the production program or due to customer input
- Employees absent due to illness, or
- Even very indirect factors such as environmental influences, for example

As this list should make clear, there is no doubt that in real production numerous things occur which will result in deviations from the "planning". This in turn makes it necessary to be making decisions almost on a permanent basis about suitable measures to take.

The type of problem the production controller must deal with essentially depends on the production sector in question, on the logistical situation and also on the vertical integration. The two examples which follow should bring this home.

With high vertical integration a two-step procedure is possible whereby operations are assigned on a central level to production islands or groups and the final decision regarding the resources to be selected and the specific sequence of production is delegated to the group.

In areas with high time pressure and short order lead times, such as is the case with automotive suppliers, planning based on strategic considerations is virtually impossible since the time between receipt of the order and its execution offers only minimal room for maneuver. A time-intensive and strategic approach to planning would here be out of all proportion to the results which could be achieved, which in most cases would only be valid for a short time anyway. To sum up, the core tasks of detailed planning and control in an integrated production environment are as follows:

- Resource allocation;
- Flow control;
- Resolution of conflicts;
- Balancing against rough planning.

7.2 Use of MES for detailed planning and control

7.2.1 Overview

From the point of view of an MES, detailed planning or, put more precisely, technical resource allocation and feedback from the progress of the real events is to be considered only as a unit. Whilst in conventional planning systems such as, for example, ERP/PPS systems, this control loop is mostly used in replanning, in current MES systems the planned world melds with the reality in the actual situation.

The core task of detailed planning is to bring the capacity-oriented orders on hand out of rough planning and into a technology-oriented allocation of available resources. Where capacity orientation in an ERP/PPS and technology orientation in an MES differ can be very nicely illustrated by the example of an injection-molding production facility which frequently has a large number of machines of the same or at least of comparable types.

Assuming that ten technically equivalent injection-molding machines are available to the planning department and work is in two 8-hour shifts. This means that rough planning has at its disposal a capacity of 320 hours per day. Differentiating the individual machines at this point – in other words, for rough planning in the ERP/PPS system – is irrelevant.

But it will be of consuming interest to production and thus, for example, to the shift foreman and ultimately to the machine operator to know on which of the machines an order will now actually be produced. What makes things more difficult is that when the ten machines are examined more closely they do differ in detail, even if this is solely to do with their current tooling states. For example, in this case only two of the machines have a certain additional unit which is required for manufacturing a small number of particular products. It may also be the case that only some of the machines – and then only in combination with particular molds – have been released for use for reasons of quality or for reasons associated with the customer.

In the case of investment-intensive installations and frequently when processes are technically complex, rough planning already mostly deals with individual physical capacities and thereby removes the need for capacity selection in the MES. In such cases the emphasis is frequently on sequence planning which takes into consideration the capacity of secondary resources.

The key requirements identified during rough planning are passed on by rough planning with the orders on hand as restrictions for detailed planning. Essentially these are the following:

- Earliest possible order start data
- Latest possible order finish date
- Order quantity

How the order is now executed within the specified limits is of no interest to the ERP/PPS level and is the responsibility of production itself. The buffer or the degrees of freedom deriving from the buffer can be used as part of detailed planning. It thus becomes clear that the handling capabilities of detailed planning are basically determined by the specified constraints. However, it is precisely in the extreme case with tight constraints

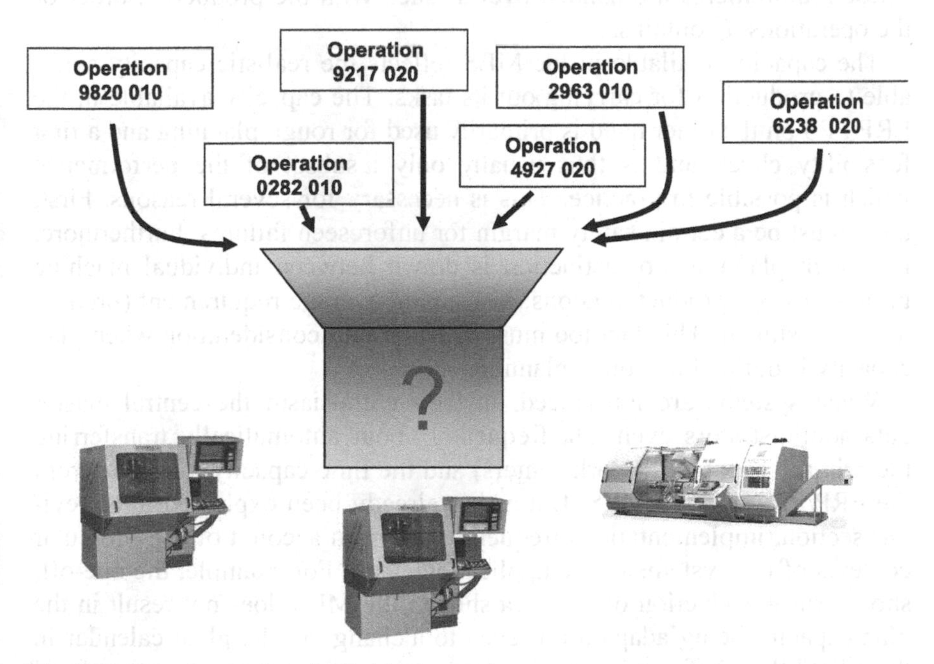

Fig. 7.1. Capacity pool vs. individual capacity

(for example, very tight target dates) that detailed planning in the MES system makes its presence felt on account of its closeness to the technology and the way the actual situation is reported back in real time.

Skillful sequencing thus allows massive reductions in set-up times or even prevents them, thereby making up for bottlenecks elsewhere. On the other hand, times required for transporting or converting supplementary modules or tools can be optimized provided their allocation is mapped as part of detailed planning, taking into consideration the real capacity, the locations where they are currently stored or installed, and also, if applicable, scheduled activities such as maintenance.

Alongside the constraints arising from the rough planning, useful detailed planning in the MES and execution of the corresponding tasks to be performed here calls for additional structures which in most cases have a technological background. These will be examined in the following sections and the need for them clarified.

7.2.2 Dealing with primary capacities in MES

One essential basic item of information is the definition of the primary capacities in the MES system. These derive from the machines or general work centers which cover the primary requirements of the operations. These requirements are handed over as such with the production order or the operations it contains.

The capacity available in the MES reflects the realistic capacity available to production for carrying out its tasks. The capacity available in the ERP/PPS on the other hand is primarily used for rough planning and a first feasibility check and is thus usually only a subset of the performance which is possible in practice. This is necessary for several reasons. First, there must be a certain safety margin for unforeseen failures. Furthermore, for rough planning no distinction is drawn between individual machine groups or even product versions; instead, an average requirement (product mix) is assumed. This fact too must be taken into consideration when plan capacity is defined for rough planning.

When systems are introduced, in their enthusiasm the central master data administrators even talk frequently about automatically transferring the primary capacities (work centers) and the time capacity available from the ERP/PPS into the MES. But as has already been explained in a previous section, implementations frequently differ on account of the particular contexts of the systems or the application cases. For example, the one-off, short-term introduction of an extra shift in the MES does not result in the plan capacity being adapted nor even to a change in the plant calendar in the ERP/PPS. Differentiation also becomes very clear in the example of specification of primary capacities. In this case, multiple physical individual capacities or machines are frequently brought together on the ERP/PPS side into a pool or capacity group. In some cases the capacities are even defined simply by mapping the cost centers.

For all types of work centers, temporal serviceability is recorded in the form of a time model or even a shift calendar. Since in the MES even very short-term changes have to be made to this model (a special shift, for example), the design of this model must be made as flexible as possible in the MES. Since it is entirely normal to make such changes and, unlike the working time calendars in the ERP/PPS, this is not done as part of the master data revision service, it must be possible to make these changes simply and intuitively, as the following example shows.

In order to be able to take special circumstances into account, different kinds of work centers are distinguished in the MES:

1. Individual/machine capacity

This is the typical type of a machine capacity which has to be allocated exclusively and which is basically defined by the time availability indicated. The availability can also be influenced by a general utilization or performance level so as to, for example, classify machines in a group which have lower performance but which can be used as technically equivalent.

Lead times are calculated exclusively on the basis of parameters which derive from the pair of work center and operation.

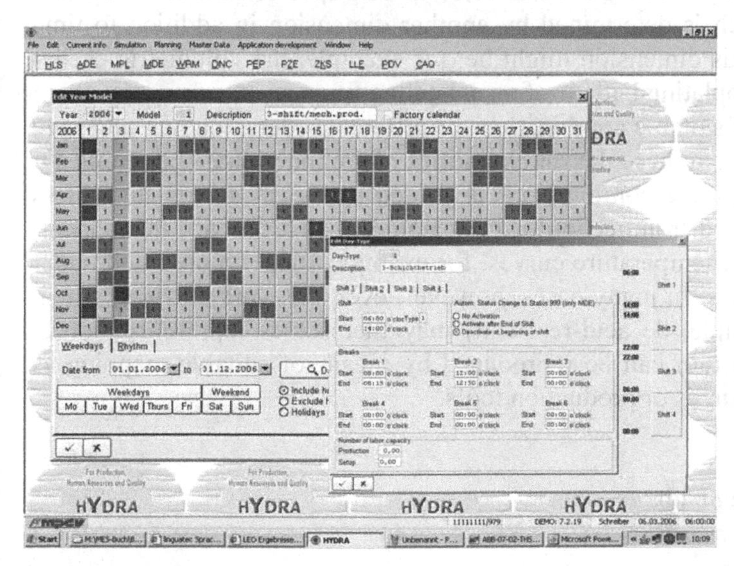

Fig. 7.2. Shift model taking MES HYDRA as an example

Examples:
- Injection-molding machine
- Screw press
- Stamping press
- Welding robot

2. Group capacity
Distinguishing it from the individual work centers just mentioned, with this kind of work center, alongside the primary capacity derived from the recorded availability of the work center itself, the specific availabilities of additional required resources are most important.

The ultimate restrictions or loading bottlenecks mostly arise from the labor capacity available or quite generally from the secondary resources and production tools required. With work centers of this kind the running time of operations is frequently guided by the type or number of resources used while the work center itself remains in the background. This becomes clear when labor-intensive processing steps are considered.

Examples:
- Assembly facilities
- Manual work centers

3. Machine capacity with second dimension (ovens, electroplating baths, and so on)
This group of capacities is to be understood as a special variant of machine capacities which is determined by another dimension in addition to time availability. This dimension might be, for example, the volumetric capacity of an electroplating bath or of an annealing furnace.

By means of suitable combinations of different orders, loading plans are obtained, and from these the running times of the individual operations. Of course, finding these combinations is guided by clearly formulated ruless which are derived from technological parameters. This means that annealing programs or temperature curves, for example, must be observed under all circumstances. In many cases, however, even the work carriers (which must, for example, be acid-resistant) may themselves represent a bottleneck, which in turn can be reproduced by means of suitable mapping as secondary resources or production tools.

Examples:
- Electroplating baths
- Annealing furnaces
- Painting lines

4. Machining centers

The machining center represents a special challenge to system-supported load planning. Here it is assumed that at a particular point in time precisely one workpiece is being machined and thus one operation involved, but that work pieces can be supplied in parallel. By suitably combining several operations and using the corresponding tools the aim can be pursued of keeping the spindles running as continuously as possible. A closer consideration of the degrees of freedom and the constraints which apply makes the scale of the task clear. Accordingly the combinations of operations should be checked by means of suitable technical parameters such as, for example, the NC program, and the individual machining times for a workpiece leveled against the loading times. Furthermore, several of these machining centers are usually looked after by one employee which means that one loading must suffice for a specific time without any other person being involved. In addition to the actual machining centers, the various tools will also have to be considered. These are often supplied from automated tool feeders which sometimes look after several installations. This results in an additional restriction.

Further to the actual load planning, a suitable running time calculation for the various operations must be carried out for this machining center. With this calculation it must be remembered that the operations, although they are active in parallel in the installation, are still processed sequentially. This means that the machine combination, the tool-changing time, the pallet capacity and the frequency of repetition will have an effect on the running time of the individual operations.

Example:

- CNC pallet machines

7.2.3 Modeling the processes in the MES

Process-capable mapping of production sequences starts with the quality of the master data and their closeness to the process. This is because on this basis calculations are made, prices are obtained and ultimately the requirements defined for implementation in actual production.

Thanks to the growing capabilities of PPS systems and in particular the increasing number of methods of mapping the processes in the master data managed in these systems, such as operation sheets and bills of material, the above-mentioned approach is actually being implemented in a large number of industrial fields of production.

Let us also assume that this approach is recognized as effective and efforts are thus made to generate high-quality master data. Furthermore, out of this awareness comes the recognition that practical experience from the processes should be permanently analyzed and should almost inevitably be used for the permanent optimization of the input requirements and thus of the master data as well.

In this way the production orders sent from rough planning for implementation in detailed planning will reflect the reasoning and the know-how behind the input requirements and master data on the basis of which they were generated. This results in the desire or rather the necessity of taking this information into account or even enforcing its use as part of implementation in production and thus in the MES.

To show how necessary this is, we may mention here the operations involved in assembling large machines such as construction machinery. The order size is very low since we are concerned with complex, large machines manufactured customer-specifically. The individual construction machines are built in different, sequential assembly steps on the basis of the individual components and after a certain stage in assembly are regarded as individualized products.

In order that the individual components can be manufactured on a coordinated basis and the effects of the individual steps can be taken into account even in the planning, the upstream production orders (components) need to be grouped together like a subassembly and treated correspondingly in the MES.

To return to the example of assembly which upon closer examination appears in the order network "only" as an individual component. Within

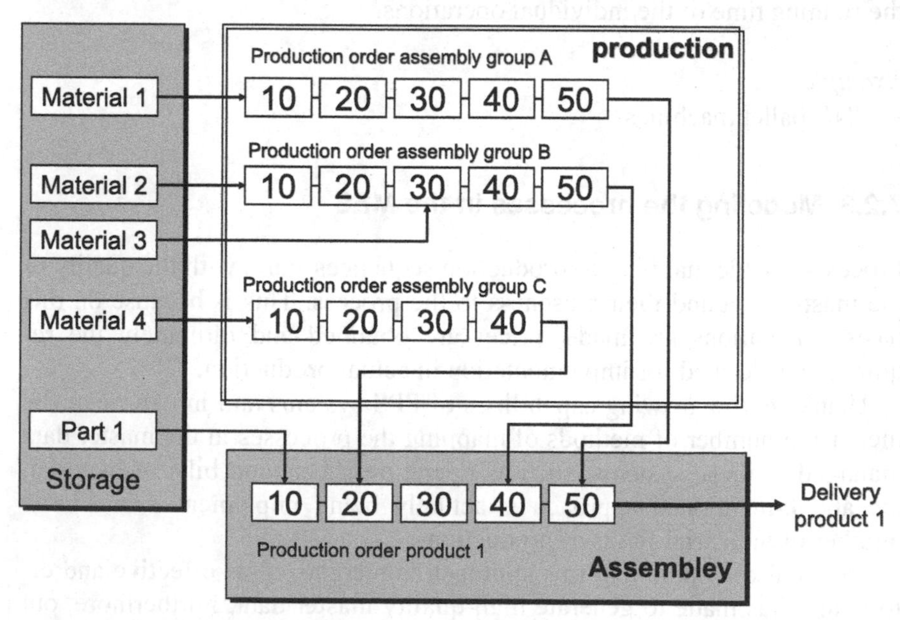

Fig. 7.3. Subassembly-based manufacturing/assembly

this individual assembly order, the sequence, in the sense of a predecessor/successor relationship, must be complied with. To ensure the flexibility of the individual work steps, it is however important that this sequence be factored in for every individualized separate machine. This is often absolutely essential in the assembly of large machines on sole account of space requirements and the intermediate storage capacity available.

Once an individual machine is in itself finished, various downstream work steps follow, such as, for example, different checking and inspection procedures, test runs or other commonplace aspects such as furnishing with documentation and other supplementary equipment.

These work steps will no longer be subject to a strict sequence but can be carried out in any order – once again with respect to individual machines. Since these are certainly time-intensive activities which also require correspondingly restrictive capacities such as a test rig, for example, very detailed planning will be necessary and thus a correspondingly finely subdivided routing plan.

The extremely complex processes occurring in the reality of this example have certainly been greatly simplified but it does very nicely demonstrate a large number of different relationships which absolutely have to be included in the detailed planning and execution of production if the processes are to be reproduced in reality. Similar scenarios can be set up for any processes or sequences and will thus come up against numerous relationships which are indispensable to an adequate image within the context of master data modeling or MES detailed planning.

In order to be able to attain sufficiently precise and realistic modeling of the processes, modern ERP/PPS systems – and thus MES systems as well – basically map the following routing plan structures:

- Orders of different order types;
- Operations for mapping technology-oriented work steps or process steps;
- Operation splits in order to make use of parallel production possibilities, such as installations of the same type, with the aim of reducing lead times;
- Single-part analysis (serialization or individualization) for examining the individual process steps with respect to each individual part;
- Order networks for networking several orders together and thus, for example, for mapping projects or a typical assembly of a subassembly;
- Variable relationships between individual or even several work steps (operations) in an order to map for example:
  - Strict predecessor/successor relationships;
  - Exact synchronization of a parallelism in order, for example, to schedule a first-off inspection exactly related to the production operation producing it.

- Mapping of subnetworks within orders in order to be able to parallelize them and if necessary reconverge them later.
- Support for overlapping in different versions:
  · CAN overlapping
  Overlappings of this kind are not included initially but can be implemented if there is a need to reduce lead times.
  · SHOULD overlapping
  Here overlapping is assumed as early as scheduling and even detailed planning – in other words, it is used for reducing the process duration. If for specific reasons overlapping in production is not possible, this is also permitted.
  · MUST overlapping
  In the event of a defined MUST overlap, this must be complied with for process-related reasons and must accordingly be complied with during detailed planning and a plausibility check carried out as part of the implementation which follows.

7.2.4 Personnel: the especially valuable resource

With the increasing complexity of production processes and the growing importance of the employee in manufacturing companies, the influence of personnel availability on production planning and conversely that of production planning on personnel organization is becoming greater.

In many companies, production planning which does not take the shift plan into account has become inconceivable. In most cases today this is still done with the help of manually drawn-up lists or the unspeakable Excel lists which many companies find virtually indispensable. This state of things is difficult to understand in the light of the functionalities offered by current and integrated MES systems, and workforce attendance and absence planning together with up-to-date attendance time recording are some of the basic disciplines which will be examined in detail in this chapter.

7.2.5 Modeling technological aspects

Alongside the relationships which arise from the bills of material and operation sheets, the process is determined by numerous technical parameters, especially those close to the process. The more detailed the consideration given to the influences on the processes the more realistic a plan will be and the more likely it will be implemented.

Technological parameters of this kind might be, for example, possible combinations of different resources in order to manufacture a particular article, such as combinations of machine and tool. Restrictions of this kind arise for technical or even for quality-related reasons. This means that in the first case one particular tool for manufacturing a product can only be used in certain installations. In the other case – irrespective of the purely technical possibility – a specific combination does not deliver satisfactory quality or quite simply the combination has not yet been approved by QA or the customer. This simple example can be applied freely to other resources such as inspection, measuring and test equipment, NC programs and even personnel, and a large number of application cases can therefore be found in practice.

Mapping an image of these circumstances means that the possibilities for covering the needs arising from the orders or operations can be specified more precisely within the context of detailed planning. Possibilities for time variation on the possible capacities emerge from the relationships (described further above) within the orders or between orders.

Furthermore, the sequence in place on a capacity-holder yields further indications which, within the context of detailed planning, will be of interest to a final decision regarding capacity loading. For the sake of simplicity let us take a machine to which the same kind of operations from different orders are to be assigned. Here not only are the time restrictions arising from the orders important but also the sequence itself since this could help prevent what might be enormous set-up times. Looked at differently, a sequence which must be complied with for reasons of scheduling can result in unnecessary conversion or setting-up costs which were not included as such in the purely product-oriented operation sheets and thus not in cost estimating either.

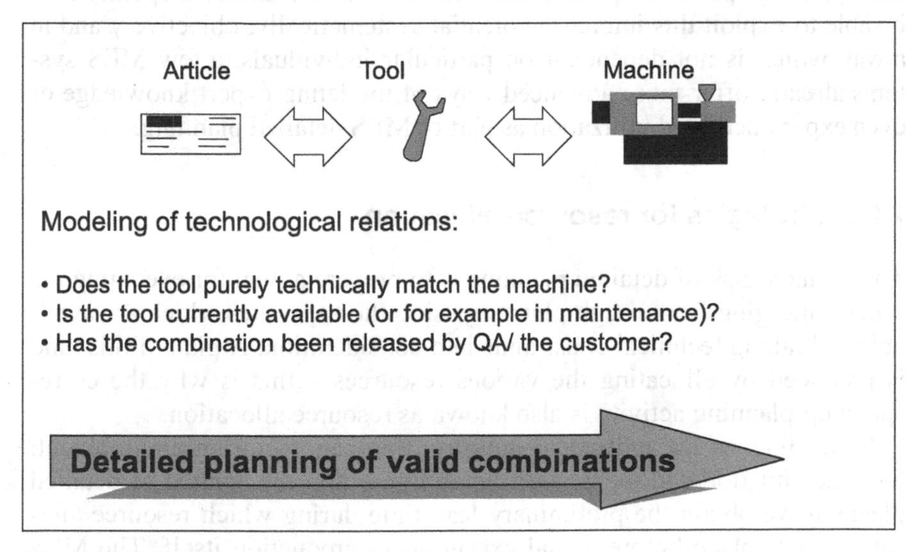

Article
Tool
Machine
Modeling of technological relations:
• Does the tool purely technically match the machine?
• Is the tool currently available (or for example in maintenance)?
• Has the combination been released by QA/ the customer?
Detailed planning of valid combinations

Fig. 7.4. Modeling technological relationships

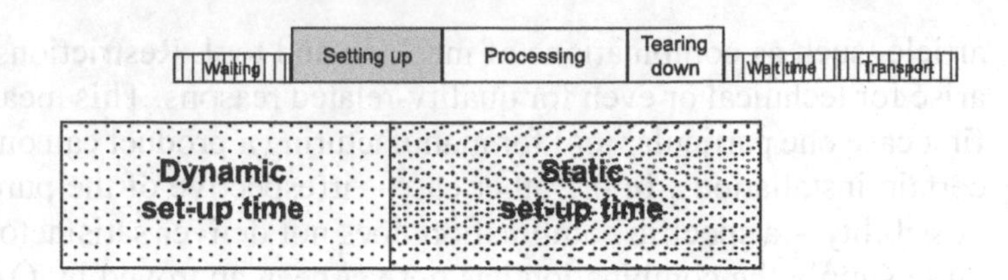

The dynamic set-up time...

- results from the setup change matrix
- is determined due to the occupancy order of a work-station

The static set-up time...

- is defined in the work schedule
- is independent of the occupancy order

Fig. 7.5. Static vs. dynamic set-up times

To allow times of this kind to be included in detailed planning, MES systems provide the corresponding structures, such as, for example, the so-called tool set-up matrix.

Beyond the aforementioned and fairly discrete parameters, there is still a large number of technological parameters which are unfortunately only to be found inside the heads of those familiar with the process, such as, for example, the operations planner and scheduler or the machine operator. To be able to exploit this immense potential systematically, objectively and in a way which is not dependent on particular individuals, a few MES systems already offer more advanced ways of modeling expert knowledge or even experiences and utilization as part of MES detailed planning.

7.2.6 Strategies for resource allocation

The primary task of detailed planning is to prepare a plan for executing the orders emerging from rough planning using the capacities which are available and taking technical constraints into consideration. A plan of this kind is produced by allocating the various resources – this is why the corresponding planning activity is also known as resource allocation.

Depending on the individual industry, the manufacturing environment, the time situation and the desired flexibility within the context of detailed planning we obtain the preliminary lead time during which resource allocation takes place before actual execution in production itself. The MES

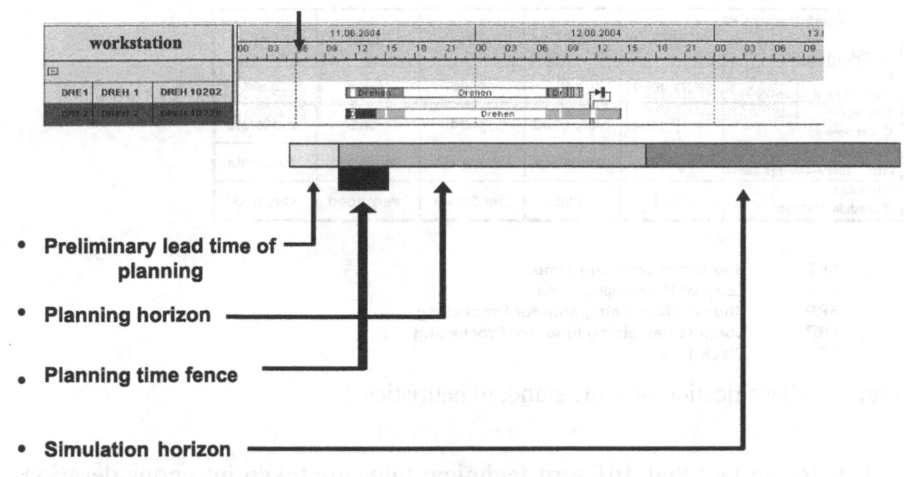

Fig. 7.6. Detailed-planning time spans in an MES

system provides different ways of displaying the time situation. Different time ranges or even time horizons are available which can be flexibly adjusted and the detailed planning be thus adapted to the specific requirements of the company.

To help users in their tasks, MES systems allow them, as part of detailed planning, to have resource allocations drawn up automatically. Taking into consideration the horizons which have been defined and the numerous relationships and restrictions we described previously, the orders on hand are planned in on the basis of what capacities are available.

A distinction is usually drawn here between replanning and delta planning. In replanning, First all allocations already made are cancelled and then the entire group of orders is planned afresh. Delta planning, on the other hand, attempts to work changes in the orders into an existing allocation plan.

Stipulating the allocation for a specific time span (the so-called planning time fence), and thus keeping it binding on production, can be done with the fixing allocations function which can be automated or be carried out via manual interaction.

Within MES detailed planning, various planning strategies are available for supporting different objectives, with some of these objectives being typical of particular branches of industry. These objectives include, for example, optimization of set-up times, reduction of lead times, minimization of work in progress inventories, and observance of deadlines.

The table below provides an overview of typical problem-solving approaches and where they are used or what they aim at.

Rule/strategy → ↓Criterion	SPT	LPT	SRP	LRP	ST
Max. Utilization of capacity	very good	bad	good	good	good
Minimum Cycle time	very good	very good	good	bad	moderate
Minimum Intermediate-storage costs	good	moderate	moderate	moderate	moderate
Minimum Schedule deviation	bad	bad	moderate	very good	very good

SPT **Shortest Processing Time**
LPT **Longest Processing Time**
SRP **Shortest Remaining time for Processing**
LRP **Longest Remaining time for Processing**
ST **Slack Time**

Fig. 7.7. Classification of some standard heuristics

Due to the fact that different technical facts are taken into consideration in the MES detailed planning, the likelihood of a resource allocation thus made actually happening is very high and can also be continuously improved by optimizing and giving the parameters a higher level of detail.

For example, restrictions due to absence of machine authorizations, to currently barred tools or to vacation situations are taken into account. This of course applies not only to automatic planning or the preparation of planning suggestions but also to manual interaction.

As regards manpower planning – in this case production personnel– two basic methods can be identified in companies which are essentially determined by the environment, by the flexibility and the deployability of the employees. The major difference very nicely describes the variance which MES systems must deal with in the field of detailed planning.

- In the one case, the manpower which is available and plannable emerges from attendance /absence planning and acts as a restriction in the production planning.
- In the other extreme case, production planning indicates what correspondingly qualified personnel are needed and this is input into short-term manpower planning as a requirement which must be satisfied.

7.2.7 Conflict resolution by simulation and optimization

As has been described in the previous section, many alternatives exist for planning orders on the basis of specified strategies. The number of variations rises drastically when manual interventions are included.

By applying different strategies and making an objective appraisal of the results, various simulations can be carried out.

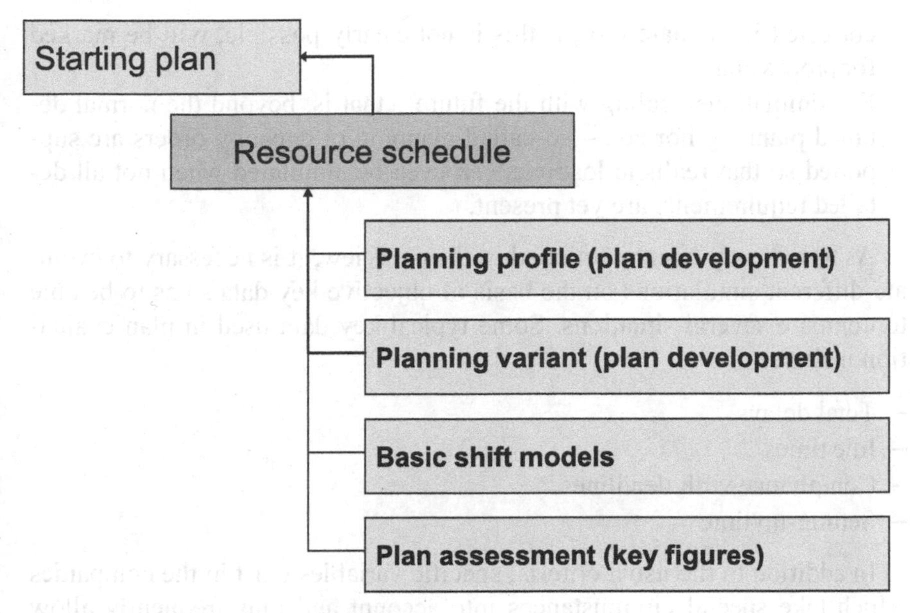

Fig. 7.8. Components of a stored plan

Due to this objective evaluation, the planning situations which emerge can be compared and better plans thus obtained.

In what follows we shall describe the basic functionalities required for carrying out useful simulations. We show the general procedure.

- It is possible to save a loading plan as a simulation. Here, in addition to the plan itself, the underlying basic data, such as the planning strategy, the capacity available and the starting situation, are also stored.
- When several simulations are run they are all produced using a single selectable starting situation as otherwise it would not be possible to compare them.
- A simulation can in turn also be used as a starting situation for other simulations.
- It is possible, on the basis of one starting situation, to automatically start several planning runs using different parameters or strategies.
- To be able to ensure objectivity in comparisons, the different plans or simulations can be evaluated objectively by means of key data and then saved in turn together with a plan, that is, a simulation.
- Once several simulations have been compared and a variant selected, that planning situation can be accepted as a binding loading plan and saved. Since in this case the starting situation will usually have changed in the meantime, this acceptance will need to take into account the current situation. Any conflicts or deviations which may occur will either be

corrected immediately or, if this is not clearly possible, will be marked for processing.

- For simulations dealing with the future – that is, beyond the normal detailed planning horizon – so-called planning or capacity orders are supported so that realistic loadings can even be simulated when not all detailed requirements are yet present.

As has already been mentioned in the overview, it is necessary to evaluate different simulations on the basis of objective key data so as to be able to compare several situations. Some typical key data used in plan evaluation include:

- Total delays
- Idle times
- Compliance with deadlines
- Setting-up time

In addition to the usual criteria, specific variables exist in the companies which take special circumstances into account and thus frequently allow the quality of a particular loading to be assessed very clearly.

To enable coverage of specific factors of this kind, user-specific evaluation criteria can be set flexibly to match the circumstances in question. As is well known, one dilemma in production planning is that the different objectives compete with each other. For example, an optimum setting-up situation

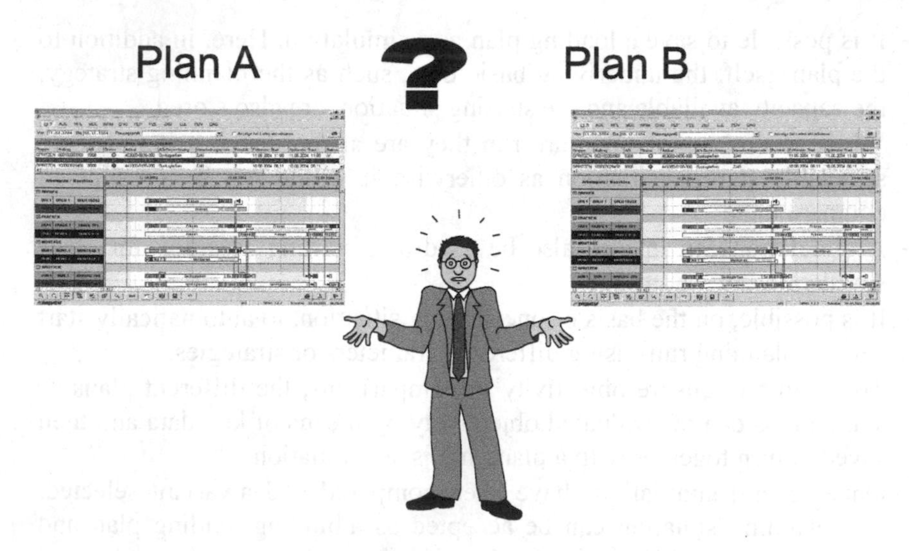

Fig. 7.9. Plan evaluation

may mean that loading or compliance with deadlines is neglected or, in the reverse case, complying with all date requests or even date promises may be at the expense of setting-up time or of the work in progress inventory.

This permanent conflict of objectives can be very clearly illustrated by the following diagram in which the corners of the pyramid represent the competing objectives and the ball in the middle the individual objective or the compromise.

By giving suitable weightings to the individual objectives they can be brought together and summarized in a combined objective – provided certain compromises are accepted. Various plans can now be generated and judged on the basis of this combined objective and thus compared with each other.

The various simulations can be generated by selecting and applying different planning strategies. If, for example, an overload situation has to be compensated for, the capacity available can easily be varied by adding an additional shift or even by simulating new installations.

If the situation as regards the individual objective can be improved by varying a criterion, the results can be further optimized by repeating the simulation on the basis of the "best" situation and by varying different parameters.

To sum up, the overall situation can be continually improved by finding "better" plans. What is of decisive importance to the quality and the success of this procedure is the coherence of the individual objective. The manufacturers of MES systems offer different strategies for system-supported optimization. Research has been going on for years into the optimization of

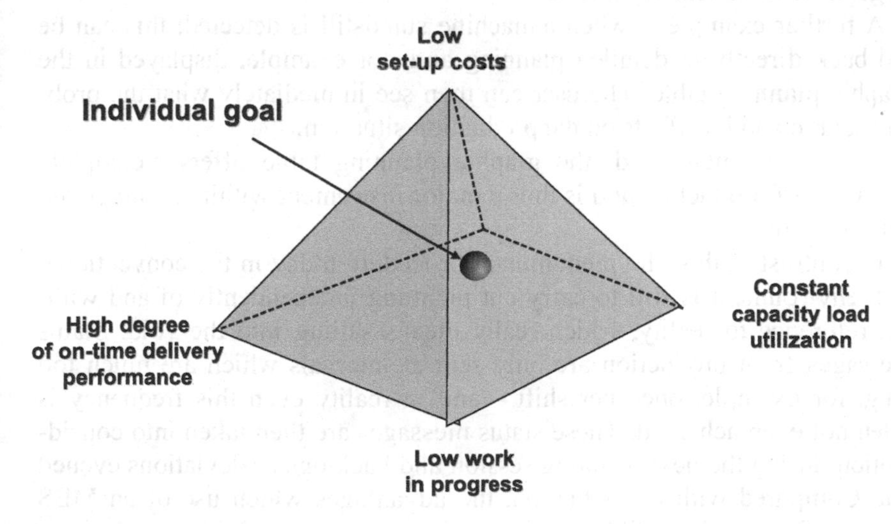

Fig. 7.10. Conflicting objectives in production planning

planning problems in general and its application in production planning in particular. Many years ago, for example, so-called optimal planning strategies were modern but due to long running times and the extreme effects on the result which arose from minor discrepancies in modeling they did not gain widespread acceptance. Current research work is concerned with algorithms which look to the natural world and which adapt like evolutionary processes. The disadvantage of such approaches is that they are extremely solution-specific and thus in their unchanged form only to a limited extent suitable for use in standard systems.

Some MES vendors have recognized this tendency and thanks to early cooperation and collaboration with institutes, universities and customers have been able to utilize their findings and incorporate them in their own product as specific functionalities.

7.2.8 Monitoring order execution

In addition to future planning, one essential task is to monitor the execution of an order. In MES systems this is done by permanent feedback of the actual situation to permit comparison with the detailed planning.

Accordingly, by extrapolating, for example, the running time remaining for an operation on the basis of the current status messages it can be seen directly whether planning can be complied with or whether, for example, the required time will not be met due to technical problems. Thanks to such information being available with little time delay, the appropriate steps can be taken or problems even be sorted out sufficiently to allow the target time to be reached after all.

A further example is when a machine standstill is detected: this can be fed back directly to detailed planning and, for example, displayed in the graphic planning table. The user can then see immediately what the problem source and its effects on the production situation.

As already mentioned, the graphic planning table offers a complete overview of production and is thus a major instrument within an integrated MES system.

In contrast to this, the predominant procedure today in the conventional PPS environment is still to carry out planning independently of and without reference to reality, which really means sailing into the void. Status messages from production are only sent at intervals which are much too long, for example, once per shift – and in reality even this frequency is often not even achieved. These status messages are then taken into consideration during the next planning session and backlogs or deviations evened out. Compared with this situation, the advantages which use of an MES system offers are incredible.

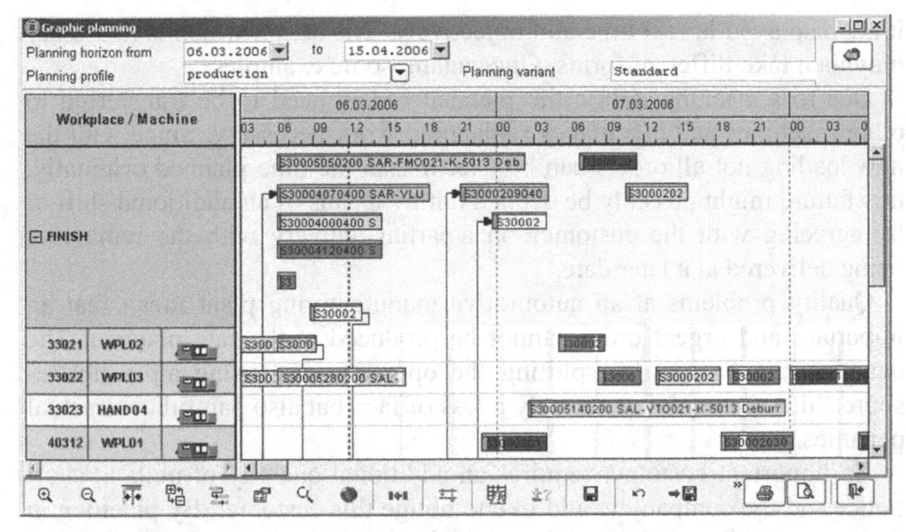

Fig. 7.11. Example of a graphic planning table used in MES HYDRA

7.2.9 Reactive planning with MES

As was already stated at the beginning, real-time feedback from production combined with the corresponding data acquisition infrastructure in the MES system means that deviations can be detected immediately and the corresponding reactive action taken. This is why in detailed planning in the MES system we also speak of reactive planning.

In addition to the planning activities which have already been sufficiently described, reactive interventions are characterized by the fact that they become necessary at very short notice and what is more as the result of an unforeseen event.

Let us provide some examples of this kind of unforeseen event with which users are routinely confronted every day.

- Technical malfunctions in machines
- Sickness of employees
- Difficulties with a material being used
- Short-notice changes to target dates or order quantities
- Tool defects

The various faults have more or less drastic effects on the environment and the planners reaction will in each case depend on the problem, its effects and above all his possibilities available. It is, however, of decisive importance that the fault itself and the systematically detectable effects of it are displayed in real time and objectively. The action which is necessary may then take different forms. Once again, some examples:

Due to a machine failure the planned orders need to be transferred to other machines and this makes replanning work necessary. Since with the new loading not all orders can be executed in the time planned originally, this failure might possibly be evened out by means of an additional shift or by agreeing with the customer on a partial delivery with the remainder being delivered at a later date.

Quality problems at an automotive manufacturing plant mean that an important and urgent order cannot be produced at the rate planned. The deadline can be met by splitting the operation and using a parallel resource, thereby avoiding not only a loss of face but also painful contractual penalties.

An important customer requires an additional quantity of units at short notice and the company would like to oblige this customer. By planning in the corresponding order and in particular by taking all technological aspects into consideration, the effects of doing so become transparent. The various faults and effects are appropriately escalated by the MES and communicated to the user. This, for example, might be a direct visualization in the graphic planning table. To be able during the night shift, for example, to notify persons on duty that a machine standstill or some other serious deviation has occurred, not only can the corresponding media be used, such as messaging on a cell phone, but an email could even be sent.

In addition to quick-response displays of faults and conflicts, the MES can support the user by displaying and evaluating what alternative possibilities exist. In a case of this kind, various scenarios can be simulated and the corresponding outcomes evaluated. This thus provides the basis for objective decision-making when, for example, a special shift or a weekend shift is to be included. It is also conceivable that expensive overtime will only ameliorate the effects to a trivial extent and thus make it unreasonable from the economic point of view. To follow this line of thought a little further, this inefficient special shift – if seen solely from the monetary point of view – can be used for special orders in order to minimize loss of face.

7.3 Management of the means of production (resources)

As should be clear from what we have said, the planning of so-called secondary capacities or even production tools is an important task of detailed planning in an MES system. We will now illustrate this necessity by means of some more examples.

Due to the increasing complexity of production processes and due also to technological process depth there is an increasing need for additional high-quality equipment and tools which are therefore only available to a limited extent. In many production areas these machine groups thus represent the bottleneck in the manufacturing process.

In ERP/PPS systems the design of the operation sheets and of the work centers is mostly based on cost-accounting or implicit costing units while the reality of production is concerned with technology-oriented units. What then occasionally happens is that rough planning considers capacities which count as non-critical as far as implementation in production is concerned while the real bottlenecks are not taken into account.

In the case of tightly interlocking supply chains even the transportation units frequently represent a limited capacity and are just as important as the means of production themselves in the preparation of a realistically feasible plan. This means that production of an article will not be possible in practice if the containers especially required for this material are not available because they have, for example, not yet been cleaned or are still in the freight forwarder's warehouse.

In sensitive fields such as, for example, the food or pharmaceutical industries, so-called logbooks must be kept for the manufacturing installations and even for all resources involved, such as mixing tanks, transportation containers or pumps, and the corresponding cleaning cycles also complied with. In order to keep up with the enormous pressure of international competition in these fields and still be able to meet very strict requirements, manufacturers are forced to map aspects of this kind by introducing MES systems.

Realistic planning of production tools is not possible without a suitable management of precisely these kinds of resources. In an MES system this task is handled by resource management which is of course not just concerned with the production tools but naturally also with the primary capacities – in other words, the machines.

Since the resources are taken into consideration in detailed planning it can be clearly deduced what resources are needed at what time at which work centers or machines. From this, MES resource management in turn generates the corresponding transportation or staging orders and ensures provision at the time required.

Taking transportation capacities into account, within the context of detailed planning these transportation jobs can also be treated like production orders and the corresponding load planning carried out.

7.3.1 Status management

The status of each individual resource is recorded, in the same way as with machines. To make things clearer, here is an example of a status model for a particular kind of resource.

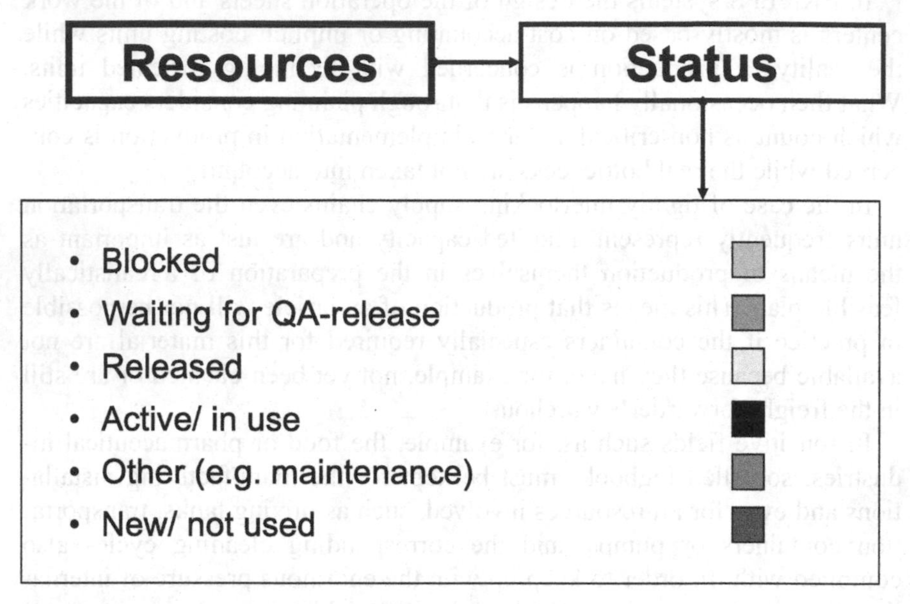

Fig. 7.12. Example of a status model

Since there are various kinds of resources, the design of the status models is also kept flexible in resource management so that different special aspects can be covered. Alongside the discrete status of the resources the storage location of the resources can also be managed.

Thanks to the integrated approach of an MES system, the quantities and times relating to processing can be posted directly to the resource. This means that not only is this information registered but we also obtain the basic data for monitoring maintenance intervals controlled on the basis of work cycles or service time. These maintenance intervals are planned in the MES's own maintenance calendar which rounds off the range of functions in an integrated MES system.

By registering and recording all activities relating to each individual resource a history is built up in the MES in the form of a log which can subsequently be evaluated by resource and thus by machine. A log of this kind is mandatory, for example, in the food industry and above all in the pharmaceutical industry.

7.3.2 Anonymous and individualized resources

Within the context of an MES system, resources are regarded as non-consumable operating resources available in a specific number and which are required for production and which do not go directly into the product. They differ in their function and use in their classification into resource types.

A *machine* is a resource assigned to a work center on a one to one basis and is thus identical to a *work center* or is a special instance of a work center. This means the machine as work center represents the primary capacity which is planned but is also at the same time a resource with the tasks bound up with that. In the MES system a machine or a work center has a unique identity.

Tools, transportation containers, process materials and *installations* are resources which in principle can be assigned to the work centers on a temporary or a permanent basis. They can occur in multiple instances and be controlled anonymously or have their own unique identities. Capacities are planned on a secondary basis. Resources are allocated either as production tools in the work plan or permanently to work centers directly in the MES system.

Figure 7.13 shows by way of example different resources which can be used in a manufacturing process:

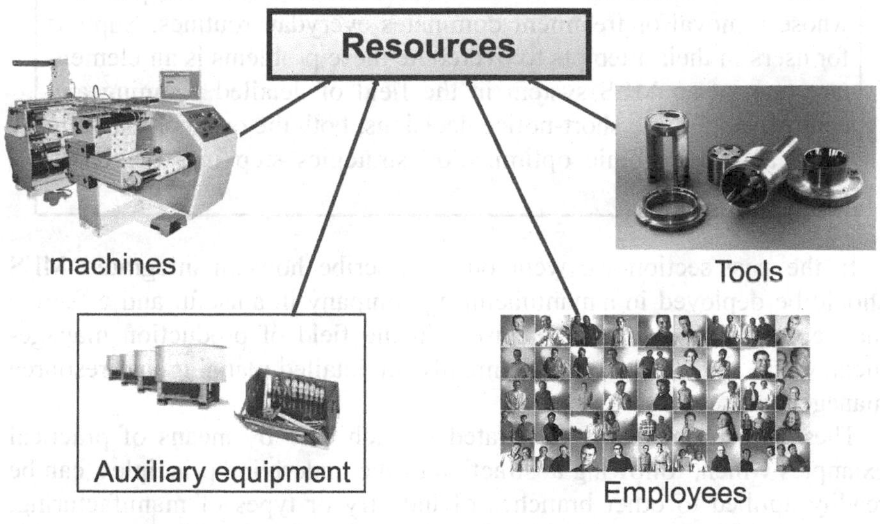

Fig. 7.13. Examples of resource types

Resources differ in their properties as shown in their classification into resource types. Tools, for example, are grouped together. In addition to its purely classificatory function, certain functionalities in the MES are also controlled by means of this grouping. For example, load planning may make sense for one type of resource but with another kind it might be that service times need to be recorded.

7.4 Summary

First of all we examined how production management and the detailed planning of an MES are integrated into the overall planning processes in a manufacturing company. The different planning levels were presented and their main focuses as regards detailing and time reference were examined.

On the basis of this introduction we went on to examine the central subject, namely the level of detailed planning and also the requirements which arise from the special task demands in this field. From this in turn emerges, as a logical consequence of an integrated production management, the management of resources on the individualized and technological level.

> The essence of this is that in the real world of production, events are permanently occurring which result in deviations from the plan and whose removal or treatment dominates everyday routines. Support for users in their attempts to overcome these problems is an elementary task of an MES system in the field of detailed planning and control. With these short-notice decisions, both the original planning goals and algorithmic optimization strategies step into the background.

In the past section we went on to describe how an integrated MES should be deployed in a manufacturing company in a useful and effective manner to provide support precisely in the field of production management. Once again the focus was mainly on detailed planning and resource management.

These approaches were illustrated in each case by means of practical examples which, following abstraction of the underlying principles, can be readily applied to other branches of industry or types of manufacturing, thereby making the specific benefits of an MES almost self-evident at this point.

To conclude this chapter we might also mention the following basic principle which every company should observe, irrespective of its sector of industry or even of its type of organization.

> The precision or level of detail in a plan and thus the effort required to draw up the plan must be commensurate with the likelihood of the plan actually being executed!

8 Quality assurance with MES

8.1 Quality in practice

Quality assurance has been and today still is an independent branch in many manufacturing companies. The historically caused division between quality assurance and production management has often resulted in an inhomogeneous system landscape. The consequence is separate message dialogs such as, for example, in production data acquisition and in-production inspection. Production orders and inspection orders are in separate systems and not infrequently defects and scrap recording is carried out in both systems. What should also be prevented is the user being unnecessarily confronted with two systems, especially as production data are also quality data. Not only that but integrating two different systems is only possible with a great deal of effort and expense. Monolithic standard products may be inexpensive but they do have clearly recognizable limits. Integration of two systems is only the case at defined locations and subsequent expansions are bound up with considerable costs.

In an MES system quality assurance is integrated into production management. The result is a reduction in message dialogs, interfaces are avoided and there is greater acceptance on the part of the users. A further advantage of integration within an MES system emerges in auditing and certification activities. Then there is also, particularly in the food and pharmaceutical industries, the requirement for FDA conformity. As regards meeting FDA requirements, best use can be made of the synergies arising from integration within an MES. Ideally these requirements will be satisfied by means of the basic functions provided by leading MES systems. In what follows we shall be describing the functionalities and benefits of an integrated quality assurance system. We will deliberately omit any description of the various standards in this area (QS 9000, TS 16949, VDA, and so on).

In addition to quality planning, methods for preventing defects and securing product quality will also be considered. In this way the completeness of the corresponding quality database can be checked as early as the in-production inspection stage. The appropriate action can then be taken to correct or remove deficiencies promptly. Further aspects of integrated quality assurance are its component areas of documentation, evaluation and analysis. Documentation absolutely does not mean the management of forms but rather the gapless and efficient presentation of all data relevant to quality. Great importance should be attached to traceability (tracing and tracking lots, batches and products). The entire product creation process can be traced from incoming goods, via the corresponding intermediate products or semi-finished articles as far as shipping of the final products.

It is not until an MES is installed that higher-level evaluations and analyses with a much higher information content become possible. Inclusion of all information relevant to production – and this in particular means machine and process data – makes it possible for efficient measures for the prevention of non-conformances and for process optimization to be introduced.

8.2 Planned quality

A quality plan is a form of project planning which aims not only to define process flows, which are oriented by corporate goals and by what customers want, but also to monitor compliance. In connection with quality planning, Sect. 5.5.2 of DIN ISO/CD2 9001:2000 states: "The organization must determine and plan the activities and means of reaching quality objectives. Planning must be reconcilable with other requirements of the QM system. Planning must cover the following areas:

- Processes required in the QM system
- Required (product) realization processes and means
- Definition of quality characteristics on different levels so as to secure the desired results
- Verification activities

Acceptance criteria and required quality records. The planning must ensure that organizational changes are carried out under guidance and that operation of the QM system is maintained during these changes." A quality planning which is transparent and complete as far as its contents are concerned lays the foundation stone for demonstrating to the customer that his suppliers are meeting requirements relating to function and to quality. An MES supports the user in the systematic early preparation and planning of all measures necessary for attaining a level of performance which satisfies the customer.

8.2.1 Quality master data in an MES

For the planning and execution of quality-assurance measures an MES system provides functions for managing basic data. The following master data are normally used for defect analysis:

- Defect types
- Defect locations
- Defect causes
- Causative agent
- Measures taken
- Cost types

These data should be held in a hierarchically organized structure. This will allow, for example, detailing of evaluations on the highest group level, starting with key defect areas.

If the defect analysis criteria are to be available in detailed form and thus in a correspondingly large number, the MES will assist the user to link just the relevant subsets of these basic data to articles, article groups, characteristics or inspection and test plans. This will help prevent incorrect inputs and increase acceptance at the data acquisition stations.

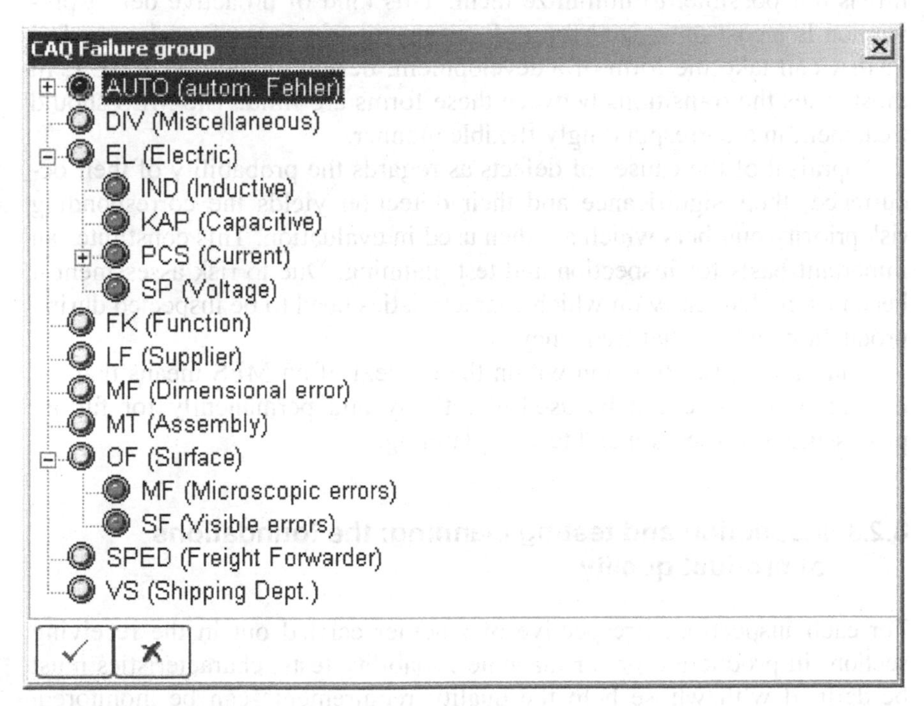

Fig. 8.1. Defect type grouping

When an MES system is deployed, one result, even on the level of the basic data, is the emergence of synergies. There is now no need for duplicated maintenance of the same kind of data as is the case with monolithic standard products. The master data interfaces they require are in most cases bound up with high costs and administrative outlay.

The following are some of the basis data relevant to quality which can be managed centrally in the MES (in other words, for all MES functions collectively):

- Units
- Work centers and inspection stations
- Defect types and reasons for rejection
- Cost types
- Persons

8.2.2 Proactive defect prevention with FMEA

With the aid of FMEA potential defects can be investigated as early as when the product itself or the production process is being designed. Suitable steps should be then taken to eliminate these potential defects or, if this is not possible, to minimize them. This kind of proactive defect prevention is a cost-optimized form of getting rid of points of weakness. The FMEA can take the form of a development, design, or process FMEA. In most cases the transitions between these forms are fluid. The MES should treat them in a correspondingly flexible manner.

Appraisal of the causes of defects as regards the probability of their occurrence, their significance and their detection yields the corresponding risk priority numbers which are then used in evaluation. This constitutes an important basis for inspection and test planning. Due to risk assessment it becomes evident early on which characteristics need to be inspected during production and at what frequency.

Using an FMEA function within the context of an MES means that the data recorded here can be used effectively and permanently for further processing in inspection and testing planning.

8.2.3 Inspection and testing planning: the foundations of product quality

For each inspection, irrespective of whether carried out in the receiving section, in production or for machine capability tests, characteristics must be defined with whose help the quality requirements can be monitored.

Equipment, activities and inspections must be established for each characteristic with the aid of specifications.

These characteristics are defined and summarized in the inspection and testing planning. Since process characteristics and product characteristics can be used simultaneously, the MES system has at its disposal all of the quality-related data it needs for analyses, certificates and control loops. With the MES, inspection and test plans can be prepared which are valid for articles, article groups, operations, customers, suppliers, standards and/or processes, depending on requirements.

Even as early as the design process, it becomes clear which characteristics of the product are relevant to quality. When an FMEA is used, a corresponding weighting can be obtained from the risk priority numbers. The integration of the FMEA into the MES allows defined characteristics to be taken over directly here. Alternatively, characteristics can also be transferred by reading in data from CAD drawings. This avoids duplicated data input, liable as this is to error.

Assignment of the inspection equipment to be used (or inspection equipment groups) makes it possible to control which measuring equipment can be used for quality data acquisition. Here the inspector is assisted by the measuring instruments and measuring equipment being connected directly. This saves time, prevents incorrect inputs and increases acceptance at the data acquisition stations.

The use of the MES system means that all product-specific data (work plans, etc.) are already available in in-production inspection planning. These data can be accessed, for example, to create the characteristics corresponding to each operation which has a relevance to quality. By making a comparison between the work plans and the inspection plans the MES is able at an early stage to reveal deficiencies in the inspection and testing planning. A direct linking together of quality assurance and production planning makes available all information about the machines involved, the tools and molds used, and the individuals logged onto machines without

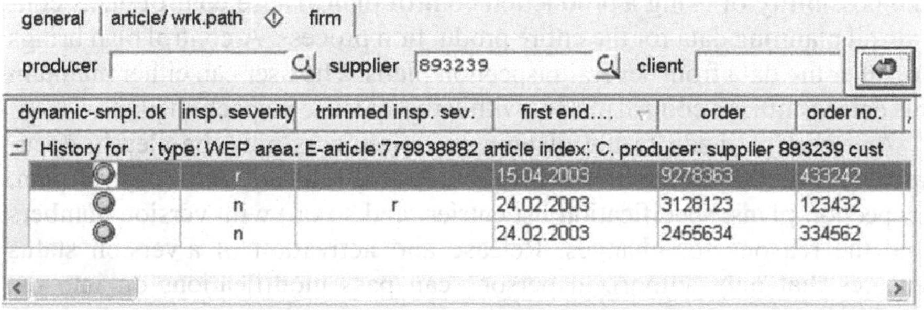

general | article/ wrk.path | firm

producer | supplier 893239 | client

dynamic-smpl. ok	insp. severity	trimmed insp. sev.	first end....	order	order no.
History for : type: WEP area: E-article:779938882 article index: C. producer: supplier 893239 cust					
	r		15.04.2003	9278363	433242
	n	r	24.02.2003	3128123	123432
	n		24.02.2003	2455634	334562

Fig. 8.2. Dynamization history

anything having needed to be done separately regarding data input or interfacing. With monolithic standard systems these data may under certain circumstances have to be maintained twice.

The production of variants makes special demands on inspection and testing planning. Here the same type of products are being manufactured and only differ in details. If every product has to be looked after with separate inspection plans this would call for an enormous amount of planning. The remedy for this problem are so-called specification lists. The contents of an inspection plan can thus be limited to a list of the characteristics to be inspected without specifications actually being given. In a separate list solely the specific features of all product variants to be produced are defined. Alternatively, configuration features can be used. Here the inspection planning for each characteristic determines how the specifications are obtained from the design dimensions. The tolerance and plausibility limits of the characteristic are relatively predetermined. The order itself will thus include the target values to be used and from these the remaining specifications are calculated. From these functionalities it can be seen how an MES system supports the inspection planner. Using these planning variants reduces maintenance effort to a minimum. At the same time potential faults due to data redundancy are avoided.

When these methods are used for dynamization the inspection frequency can be markedly reduced since empirical values are available. These functions are used primarily in receiving controls. Dynamization requires planning beforehand as to what rules it should follow. Apart from the use of the usual standards (ISO 2859, ISO 3951, and so on) the MES system provides the inspection planner with the means for creating his own rules. Alternatively, a dynamicized inspection can also be used in production following the occurrence of a non-conformance, for example, to increase the inspection interval temporarily in order to see how effective a corrective measure is.

To enable him to meet the requirements for gapless documentation of all quality-related data along the full value chain, the user of an MES system has the possibility of using a production control plan. These control plans contain all planning data for the entire production process. A control plan brings together the data from several inspection plans. The user can either maintain his data within the control plan or within the specific inspection plans.

Any change made to the planning requirements must be clearly documented. For this reason the MES provides all relevant data (control plan, inspection plans, specification list entries, and so on) with version numbers and the reasons for changes. Release and activation of a version status ensures that only authorized persons can pass modifications on into the production process. Furthermore, changes can be planned at a preliminary stage and specifically activated at a fixed point in time. The version management applet allows automatic documentation of when, why and by whom changes were made. The MES keeps these data available for research purposes such as, for example, into the history of particular parts.

If the MES includes an initial sample inspection, the user receives support in importing the corresponding characteristics into production inspection plans. All relevant settings are imported and can be edited if so desired. Even this reduces the planning effort and prevents those errors which can occur when quality characteristics are copied manually.

8.2.4 Inspection equipment: reducing measurement uncertainties

Inspection equipment is subject to wear. Inspection equipment cannot be used unless it is capable – in other words, it meets manufacturer and process requirements. To secure this capability, examinations on the basis of particular standards (QS 9000, VDI 2618, and so on) must be carried out at regular intervals. What this means is that before inspection equipment is used, activities, means and dates must be defined to secure inspection equipment capability.

Use of an MES means that the possibilities of an efficient management of inspection equipment are fully exploited. As part of quality planning, definitions are drawn up for inspection equipment capability checks which lay down which characteristics should be monitored with which resources and against which specifications. Depending on which standard is being used, it must be established beforehand which statistical parameters (repeatability/measuring device scatter, comparability/test scatter, repeatability/comparability, scatter from part to part and total scatter) must be used as basis for the capability record.

Using an MES inspection equipment management applet simplifies the planning of capability investigations. In calibration planning the intervals for calibration are decided, and these can be time-based or piece-based intervals. When determining the due times by piece-based intervals, use is made of the information available to measured data collection regarding the inspection equipment being used. With the aid of a wear factor, it is possible for every work center where inspections are being carried out to define how rapidly the inspection equipment will be detrimentally affected in this environment. For example, an inspection device in a machine which is exposed to oil and coolants will wear more quickly than when the device is used in a clean room. With the aid of these functionalities the MES can considerably reduce the number of calibrations to be carried out since realistic wear-oriented time scheduling is possible. When advance warning times are used, the time buffer can be individually defined from the first notification up until the time calibration is actually due.

8.2.5 Supplier rating: optimizing the procurement process

The quality of the materials flowing into production has a great influence on the quality of the products, especially in the light of increasing specialization and the reduction in vertical integration. Before the best suppliers can be selected in the purchasing department, effective evaluation methods must be deployed. One of these methods is supplier rating.

Supplier rating is a classic domain of ERP/PPS systems. Here all relevant factors are collected with data being read in from subsystems via interfacing functions. Key data are obtained for the various suppliers from all of these individual factors and possibly too from subjective criteria as well. These key data are used for rating the suppliers and are also very useful during meetings with suppliers. These methods make it possible for the quality of supplied goods and materials to be continuously improved.

Once the MES has all of the data used in supplier rating, in straightforward cases supplier rating can even be carried out within the MES system itself. Since real-time data are available in the MES it is also possible to take a snapshot of the current quality situation of a supplier.

Before supplier rating can be carried out in the MES system, the corresponding criteria must have already been defined or taken over from the ERP/PPS system. This takes the form of rating catalogs. In these, different evaluation blocks can be defined in a freely definable hierarchy. Evaluation criteria are then assigned to these blocks, being classified at the same time into the categories of "subjective" and "automatically determinable". While the subjective criteria have to be assessed manually, the automatic criteria obtain their results from the MES datapool directly.

Both the evaluation criteria and the evaluation blocks can be weighted differently. The block is graded is on the basis of the current grading of a criterion and its weighting. The supplier rating is calculated from the evaluations and weightings of the corresponding blocks.

In the rating catalogs as well the MES system applies the versionization and activation procedures already familiar from inspection planning. This ensures that even in this area, all changes in the planning basis are gaplessly documented.

The decisive advantage in using an MES is that all of the data present in the system can be called on directly when assessing suppliers. The result of the MES supplier rating can either be used directly or be consulted within the ERP/PPS system for a comprehensive assessment of a supplier.

8.2.6 Setting up workflows with escalation scenarios

Before the quality requirements made of products can be met, it will be necessary to plan out all of the measures which assist quality assurance. These include not only managerial and operative-level activities but also the reaction to events affecting quality.

In the planning field, process sequences relevant to quality are defined in the form of workflows. Due to the scope of functions provided by the MES, the definition of these workflows can access considerably more possibilities than is the case with monolithic standard products. For example, detecting non-conformances during the downstream quality inspection can have a direct impact on production in progress. In the simplest case, foremen or shift foremen handling jobs for the nonconforming article which is being produced or is about to be produced are informed about the problems.

Complaints too can be processed more efficiently with the aid of saved workflows. The MES thereby ensures that proven process sequences are taken into account in complaint processing. Another advantage of this kind

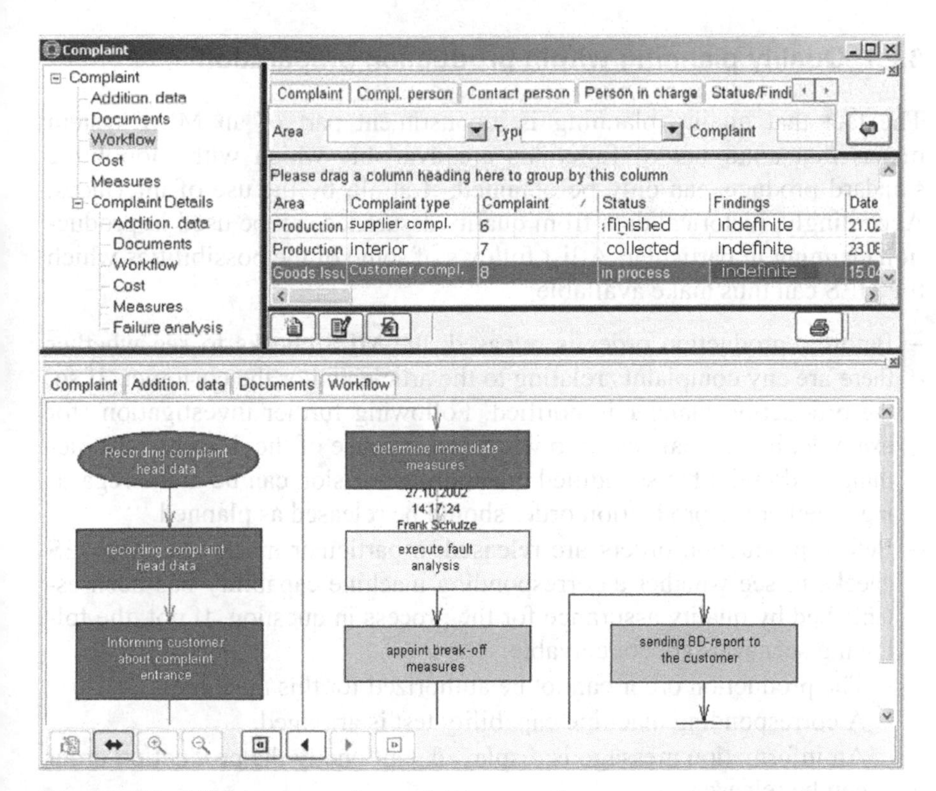

Fig. 8.3. Workflow-supported complaint processing

of structured complaint processing is that the current status and a history showing the various actions taken can be displayed at any time on the basis of the workflow steps completed.

To be able to take advantage of all of these possibilities, the underlying workflows first need to be defined. If at all possible this should be done graphically as only in this way will complex sequences remain transparent and allow necessary modifications. More advanced MES systems offer a flexibility which permits even different forms of workflows to be used for a different data context (customer complaints, internal complaints, and so on). By monitoring the workflow in conjunction with time-based control, the MES ensures that the sequence does not come to a standstill. Use can also be made of escalation management functions during planning. This means that individuals (or groups) defined subsequently can be specifically notified.

The continuous improvement in the workflows results in a marked increase in efficiency in operational processes. Flexible workflow control and monitoring with a wide variety of possibilities for interdepartmental communication is a great strength of an MES system.

8.2.7 Quality planning within production preparation

The fact that quality planning is a constituent part of an MES system means that a number of functions are available which with monolithic standard products can only be obtained, if at all, by the use of interfaces. Accordingly, historical data from quality assurance can be used in production planning in particular. A list follows of some of the possibilities which the MES can thus make available:

- Before a production order is released, the MES checks to see whether there are any complaints relating to the article and/or the customer. If so, the production planner is notified. Following further investigation (for example, by an analysis as to whether the cause of the defect has something to do with the scheduled machine) a decision can be made regarding whether the production order should be released as planned.
- Before production orders are released to particular machines, the MES checks to see whether a corresponding machine capability has been established by quality assurance for the process in question. If not, the following scenarios are conceivable:
  - The production order cannot be authorized for this machine.
  - A corresponding machine capability test is arranged.
  - An information message is displayed. Optionally the production order can be released.

- Before a production order is released, the MES checks to see whether there is an initial-sample release for the article to be manufactured for the customer. If not, the following possibilities exist:
  - The production order cannot be released.
  - The initial-sample release is obtained.
  - A corresponding initial sample inspection is arranged.
  - An information message is displayed. Nevertheless the production order can still be released.
- In production planning the MES system can access c_m and c_{mk} values from quality assurance. These data will then be available to assist in an optimum machine assignment.

An MES is superior to monolithic standard applications not only in the analysis of historical data. For example, even as early as the production planning stage, possible flaws in the inspection and testing planning can be picked up. The appropriate steps can be taken to close these gaps even before production starts. Some examples of this are listed below:

- The combination of production planning and quality planning means that inspectors and measuring equipment can be planned in. Bottlenecks are detected in good time and can be prevented by the appropriate corrective action. An additional benefit derives from the fact that the inspectors can be planned in according to their qualifications. The MES provides information and virtual assistants to help the production planner and point out potential problems.
- When it calculates the lead time for a production order the MES can access information from in-stream or downstream inspections. With this information more realistic conclusions can be drawn regarding production times.
- When production orders are released it will have been ensured that the corresponding planning level of quality assurance is present in its entirety. Deficits (missing inspection and test plans, and so on) can be corrected by timely introduction of the corresponding corrective measures.
- During production planning, the MES ascertains whether calibrations of measuring equipment assigned by inspection planning will become due during the relevant production period. This information is displayed by the MES and corresponding measures can then be taken. The resulting production delays are available beforehand as displays. This functionality deploys its full effectiveness when piece-based calibration intervals are in place.

8.3 Integrated quality

Process quality is a necessary condition of product quality. High-quality products can only be manufactured by capable, controlled processes. If a defect is detected in good time, for example, as early as the design stage, considerably lower costs will be incurred than with discovery during production, final checks or, even worse, by the customer. This is impressively illustrated by the so-called rule of ten for failure costs. According to this, the costs of failure will rise tenfold at each process step where the defect is later (with respect to its time of origin) detected and corrected. To prevent defects, quality-assurance methods must be incorporated in the entire process chain, from incoming materials via production as far as dispatching.

Capability checks carried out in production secure the suitability of production processes and the machines involved in them. This ensures that quality-related characteristics can be produced within the specified tolerances. With the aid of the MES, use can be made of the findings from these checks in statistical process control (SPC) in production.

An inspection resources management function is necessary in order to demonstrate the capability of inspection resources and to monitor information supplied by the manufacturer. It is only with the aid of regular calibrations that it can be ensured that the measured values are realistic. The MES system supports the user in the management of inspection, measuring and test equipment, the corresponding calibration and their provision to users. In addition, scheduled dates are monitored which means that notification can be given in good time when calibrations are impending.

Non-conformance management regulates the processing of complaints, defect analysis of them and the implementation and monitoring of immediate responses and corrective measures. To ensure that defects analyzed in complaints do not have any impact on current or future production processes, MES's workflow control supports the user by exerting an influence on inspection planning and on current receiving controls, production inspections and final inspections.

Provided an MES is used systematically, it will not be long before a broad knowledge base can be accessed. Here, in the event of a recurrence of a problem, the existing knowledge is used (by accessing successfully applied problem solution strategies) so as to be able to react faster and therefore less expensively. By creating quality control loops, the MES system makes a decisive contribution via defect prevention to reducing costs while improving product quality and customer satisfaction.

8.3.1 Quality via information management

The integration of production and quality management means that all relevant data can be clearly displayed within an application. Users do not need to run several systems in parallel in order to view information or maintain it. There is no need for redundant inputs and/or expensive interfaces.

When an action and escalation management function is integrated into the MES system, standardized tools become available for systematic and documented notifications and assignment of tasks. This results in short information paths between shift foremen, machine operators, inspectors and quality managers. Integrated progress control helps users not lose track of important activities.

With the aid of workflow-based process control, in-house know-how can be mapped transparently and used efficiently. Should an employee in the quality laboratory report sick, for example, his supervisor will be notified of the orders in which inspections are affected by his absence. This allows the absent employee's tasks to be reassigned in personnel planning to other employees with the same qualification.

8.3.2 Securing supplier quality

Product quality is decisively affected by the quality of supplies. If no quality agreement has been made with the supplier, receiving controls will reduce the risk of nonconforming materials or components getting into the production process. Dynamization reduces the inspection and testing outlay arising from this. With this, once there have been several defect-free deliveries of an article from a supplier, inspections are cut back. Should a defect be subsequently detected, inspections can be stepped up again.

When an MES system is used, production can access the results from receiving controls. The data relating to the particular receiving controls from which the incoming materials or components derive are made available by material and production logistics. Should problems arise during production which can be traced back to defective incoming goods the MES can take immediate steps to tighten up inspection of deliveries.

If consignments are rejected as a result of receiving controls, it is possible in the MES to draw up a supplier complaint immediately.

On the other hand, when goods are received an immediate check can be made on the basis of current and historical supplier complaints as to whether the consignment should be classed as potentially critical. A notification is issued with a view to making employees more aware of the problem.

8.3.3 In-process quality assurance

In most cases a statistical process control system is deployed to ensure that the production process is quality-capable and controlled. Alternatively, it is also possible to employ random sample testing, the so-called lot acceptance inspection by sampling. What these two methods have in common is that by means of statistical calculations they can provide information about the current quality situation of the production process. These can be used to directly influence the production process, if need be with the aid of defect analyses and the corrective and remedial measures derived from them. These inspections thus serve as a basis for control loops which represent a completed action sequence in order to generate a quality product within a process.

It is precisely in in-process inspection that an MES delivers considerable advantages in comparison with monolithic standard products. Accordingly, by means of a corresponding marking of quality-related objects (articles, operations, lots) the quality management functions and displays which are actually needed are activated. The way an MES synchronizes production management and quality management enables a unified view at the manufacturing process. This results in a plannable and transparent overall structure

Production order 34581

Operation 0100 Turning	Target quantity: 2200 Yield: 2212 Scrap: 78
ADE: Start date: 21.03.2005 Cycle time: 18,5Std Priority: 3	**CAQ:** Diameter: 20mm ± 3mm Width: 12,3cm ± 0,3cm Burr: ok / not ok

Operation 0200 Surface refinement	Target quantity: 2200 Yield: 2203 Scrap: 9
ADE: Start date: 23.03.2005 Cycle time: 103 Std. Priority: 3	

Pperation 0250 Laboratory test	Target quantity: 2200 Yield: 2198 Scrap: 5
	CAQ: Layer thickness 5µm ± 1µm carbon content 0,5% ± 0,04% Surface ok / not ok

Fig. 8.4. Jobs with production- and inspection-relevant operations

instead of the previous separation into productive operations and inspection steps. Here a production order can include several operations taking different forms.

Here the results of the quality inspection can influence following operations. For example, unsatisfactory characteristics may make a reworking step necessary. In the same way the inspection decision can have an impact on use of the semi-finished or finished products so created.

Further advantages of production management and quality management integrated by an MES system include:

- Yield and scrap quantities are reported back and evaluated in a uniform manner irrespective of whether the classification was made for production- or quality-related reasons. Individual systems continually have problems with the different handling methods.
- In all displays and reports, the order progress and the corresponding order quality can be presented simultaneously.
- In the case of production-related data acquisition at the machine all relevant data can be seen at a glance. Alongside the current quantities, information about impending inspections and about out-of-limit conditions which have occurred is also displayed. The user's attention is drawn to quality-critical information by means of color-coded displays.
- Even with interval-based inspection all order data are available directly and without detours. Direct access to production quantities means that piece-based intervals integrated into the production process are possible.
- With time-based intervals, taking machine status into consideration can result in the inspection being suspended. This means that realistic inspection intervals are obtained which would not have been possible with separate acquisition of order, machine and quality data.
- Assuming the corresponding interfacing exists, if there is a deterioration in the quality situation it can be possible to influence the machine directly. In the case of an automated quality inspection, for example, if an intervention threshold is crossed this can trigger a machine shutdown.

8.3.4 Optimization of inspection equipment monitoring

Inspection equipment management is required to ensure that measuring equipment is only used when it has been uniquely identified. Measuring equipment must be inspected and approved for use. Testing is carried out to ensure that the measuring instrument corresponds to the specifications of the manufacturer. The measured values must fall within a defined range of tolerance. The results of the test must be reproducible. If the inaccuracy of measurement determined cannot be reconciled with the quality requirements the inspection equipment in question must be barred from use.

The MES supports the user in managing and monitoring his inspection, measuring and test equipment. The fact that even these inspections take place within the MES means that all information is available regarding frequency of use. This information could, for example, be used for determining wear-dependent calibration intervals. In the case of piece-based calibration intervals the dates scheduled for calibration can be monitored in real time and directly at the machine.

Should an item of measuring equipment become due for calibration while it is in service, the inspector will be informed directly. As an option the item could be barred from further use. Integration of quality management means that measuring instruments can, if so required, be calibrated on the spot. This may make sense particularly when inspection equipment cannot be moved, as is the case with inspection machines.

8.3.5 Transparent complaints management

A complaint is always the result of poor quality, whether this be erroneous or inadequate information or unsatisfactory product quality. It does not matter whether it is an internal complaint, a customer complaint or a supplier complaint.

MES complaints management provides support in the registration and processing of complaints. Furthermore, the immediate responses and corrective measures taken are monitored. The system also assists the user in defect analysis. Naturally all of the data relating to a complaint, including the costs, will be available for subsequent evaluations.

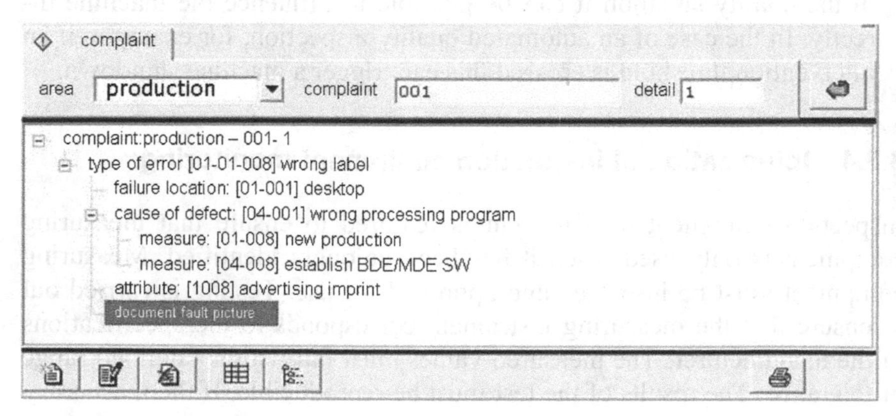

Fig. 8.5. Example of a structured defect analysis

Approved processes are taken into account by means of workflow-supported complaint processing. At the same time the processing of a complaint is gaplessly documented. The scope of functions provided by the MES results in additional synergies which monolithic standard products do not offer. Certain reasons for a machine malfunctioning will lead to an internal complaint being automatically generated. The corresponding fault analysis together with the resulting action can be used in the future for preventing problems of the same kind or at least minimizing them.

Integrated escalation management also opens up new possibilities. When a complaint is created for an article for which scheduled production orders exist, the quality officer can be notified so that suitable preventive action can be taken to avoid further complaints.

8.4 Documented quality

Appropriate handling of "information" considered as a commodity is an increasingly important factor in the competitive situation on the market. More and more companies thus regard their documentation as forming a major quality criterion for their own products. Also increasingly important is the requirement that information of decisive importance can be retrieved immediately without any great loss of time or any costs being incurred. This is of particular interest in the case of complaints, product liability issues, and meetings with suppliers or customers. The broad database in a MES which was created by the interaction of different modules allows the company to process the corresponding tasks effectively.

Interpretation of the term "documentation" is complex as regards quality assurance. Basically documentation means the efficient and gapless acquisition, management, archiving and preparation of quality-related data.

A closer examination reveals different aspects of the term "documented quality". On the one hand we have the centralized access to documents which are needed in the various quality assurance departments or which are created there. It should be noted that there is no requirement whatsoever for these documents to relate exclusively to quality. What is also important is access to information from departments outside QA/QC, departments which are "quality-remote". On the other hand, documents which are created in quality assurance, for example, as a result of recording measured values, should also be made available interdepartmentally. With an MES system, information can be mutually supplied in a suitable manner. What is of decisive importance here is the networking within an MES system (horizontal integration), since efficient data access deteriorates as the number of installed systems in which the data originated increases.

In addition, information relating direct to the quality data should also be included. The influencing factors which are significant in quality assurance documentation should be identified beforehand. Conversely the effect of changes in the actual quality data on "quality-remote" corporate and product processes must be taken into account.

Traceability assumes a particular importance here. It is important on account of product liability, and in particular the legislation which came into force in 1990 with the new reversal of the burden of proof, a duration of liability of 10 years, and an expanded coverage of the term "product". In addition there is also the August 2002 legislation concerning changes to the rules about compensatory damages according to which claims for compensation for pain and suffering now come under product liability.

Another reason for introducing a traceability management function is Regulation (EC) No. 178/2002 of the European Union which states that all companies in the food industry must implement a traceability system by 1 January 2005.

8.4.1 Networking of information

In the quality data acquisition function of an MES system not only can order documents, tool and machine information be accessed but also data relating to production planning. This means that the limitations in obtaining information found in the island solution "quality assurance" have been dismantled. The planned duration of production and thus the completion data are thus equally as accessible as information on how much of the current job has been completed. Furthermore, the inspector can examine stored work plans and thus obtain an overview of the subsequent production process. Only this comprehensive access to all relevant information will make it possible for the correct action to be taken with quality problems. In the case of areas of weakness which are due, for example, to tool wear, it can be ascertained immediately whether a replacement tool is available or whether a complete re-tooling to a different product is necessary. Global access to the information relating to the next scheduled maintenance for the machine means that a decision can be made straightaway to bring the maintenance forward. An MES system supplies all required information in the latest state required.

The MES allows any user to access the data which are relevant to him. Documents need only be prepared once, after which they only require networking with the individual function blocks. This gets rid of redundant document maintenance and assignment. For example, an article drawing can be stored centrally with the article master data and be available to all MES modules. This centrally stored document can be accessed not only during acquisition of measured values in the test laboratory but also in connection with operation and order information at the machine.

8.4.2 Using quality data appropriately

It is becoming more and more important to supply information from different divisions of the company in a clear and well-organized form and to link such information functionally. Exclusive access to quality data is no longer sufficient. When all data can be accessed, quality data can be traced back to their time of origin. Friction losses are minimized since all individuals with the corresponding entitlement can call up the information at any time and at any location. Not only the quality managers but also all other departments, from purchasing and development to servicing personnel on the customer's premises, can work using a uniform database. This ensures the necessary flexibility, especially in functional interlinking, without additional costs being incurred.

One example of this is the inclusion of the employee's "operator's license" from personnel management. While it was once sufficient to have documentation of "who inspected what", today it is becoming increasingly important to take into account the requisite inspector qualifications. Inclusion of the employee's "operator's license" means that it is possible to prevent tests or inspections being carried out by individuals who do not have the necessary qualifications. When validity criteria are taken into consideration it is even possible to check whether the inspector has inspected a particular article often enough over recent months to continue to retain his qualifications for inspecting this article. If necessary, he may need to requalify.

Classic examples of documenting quality data are the testing and inspection certificates in accordance with EN 10204, such as inspection certificate nos. 3.1 and 3.2, and specific test report nos. 2.1 and 2.2. On the part of the customer there is very often a demand for special certificates with a greater content. The form and nature of these certificates may vary from customer to customer. The relevant agreement is often limited to stipulating that in the case of a complaint, for example, a certificate has to be provided which gives the information desired. In most cases problems arise at the latest in expanding the contents to include relevant data about the corresponding process characteristics. This is not the case when an MES system is in service. The required documents can then be given a flexible design as far as the contents are concerned and can be adapted to customer requirements.

The inclusion of all quality-related events which affect the product quality represents an expansion of the classic documentation. This includes interconnecting production data and quality data. Faults which occur during the product creation process have a great influence on the resulting product quality. The same is true of the characteristic values of process, machine and tool parameters. All in all there are five groups of causes which are responsible for deviations in quality:

- The *individual* as an operator or inspector;
- The *material* used as the result of an upstream production process;
- The *machine* with the elementary influencing factor, the tool;
- The *method* which decisively determines the interplay of the individual and the machine, and
- The *milieu* with various environmental influences such as temperature and dust, for example.

One example of the importance of the "tool" as an influencing factor is the risk of voids forming should a specified mold temperature not be reached. Due to the dependence of product quality on the diversity of influencing factors associated with the cause groups mentioned above, the interlinking of all information takes on a great importance. This is made clear by the gapless documentation of part history which is required as part of product liability and which is a constituent part of an MES system.

The part history must include other information in addition to known influencing variables. It must always list all important events which occurred from the time the drawing was prepared to when the tools were scrapped. Should there be modifications, these must be described and reasons for them given. Possible contents of a part history include:

- The date of sampling and the beginning of deliveries for production purposes;
- Information about tool repairs;
- Information about process optimizations;
- Information about release changes, and
- The use of new materials.

In the case of complex products, such as machines or automobiles, in addition to the historical approach there is also the requirement that events be integrated into the part history following completion of the product. If a safety-related component is replaced, this change must be documented on account of the traceability required in connection with product liability. To enable the correct response to be made in the case of a damaging event, which person must respond to which events will be stipulated. Should the damaging event occur, progress control of all activities will be required. These in turn must be integrated into the part history.

Under no circumstances should the changes in quality-related data documented in a part history be considered in isolation. It should instead be remembered that every change can also have various impacts on other departments or processes. While there is a risk with monolithic quality assurance that these changes will remain undetected or not necessarily automatically have an effect on other departments, an MES system ensures that information is forwarded immediately. All corporate departments or divisions have direct access to the revised contents or documents. Consequential costs caused by a delayed inclusion of changed and quality-related data are avoided. Any correction of measured values which becomes necessary after the event can have an effect on the inspection decision and thus on the quality status of the corresponding production lot. In such a case the MES system prevents further processing of the rejected lot in all downstream production processes. All lots into which the rejected lot has already been packed will also be rejected and prevented from being shipped. In the worst case, products which have already been shipped may have to be recalled from circulation. To limit the damage and the associated costs the automatic functions of an MES system will be required. Since there is collective access to all relevant data a swift and effective reaction is possible.

Impermissible deviations detected during inspection equipment calibration or during machine or tool maintenance can have similar effects. The results documented with this inspection equipment may be based on incorrectly measured values. In this case as well the production lots in which the inspection equipment was used must be barred or a reworking note issued. In the event that these impermissible deviations cannot be corrected, the item of equipment must be quarantined and thus no longer be available as a resource. Upon the equipment being barred for use, the MES must check immediately to see whether it has already been planned into future production orders. If no alternative resources are available, replanning will be necessary.

Causes of defects identified even during the course of a complaint investigation can have direct effects on the production in progress or impending. If the cause of the defect is determined to be an incorrectly assigned DNC program or to maintenance which was not carried out, appropriate action should be initiated without delay. If necessary, the current machine settings and the tools being used should be checked. It is important that the checking process is instigated rapidly and that it should be automated as much as possible so that consequential costs can be avoided. The use of MES's higher-level escalation management function can also make a contribution here. Once a complaint is formally opened this function allows an automatic check to be made to see whether the corresponding article is in any scheduled or current production orders. If the check is positive, the production manager or the planning manager will be informed. If specific defect causes are assigned or actions identified, an escalation can be initiated together with automatic notification. Should receipt of the message not be acknowledged, other individuals such as deputizers will be informed automatically.

These examples illustrate the effect of changes in quality-related data or of decisions on other corporate departments or processes and reveal the necessity of deploying an MES system as an integration platform for previous island solutions.

8.4.3 Traceability

To make traceability possible, all lots or batches or even individual products must be uniquely labeled. This must be kept up throughout all stages of manufacturing and all logistical processes. Only in this way can the origin of products be determined along the value chain.

Traceability includes not only tracing but also tracking a product within the creation or delivery process and requires that the information flow be linked to the physical flow of material.

The reasons for introducing a traceability function may be statutory requirements relating to product liability or general customer requirements.

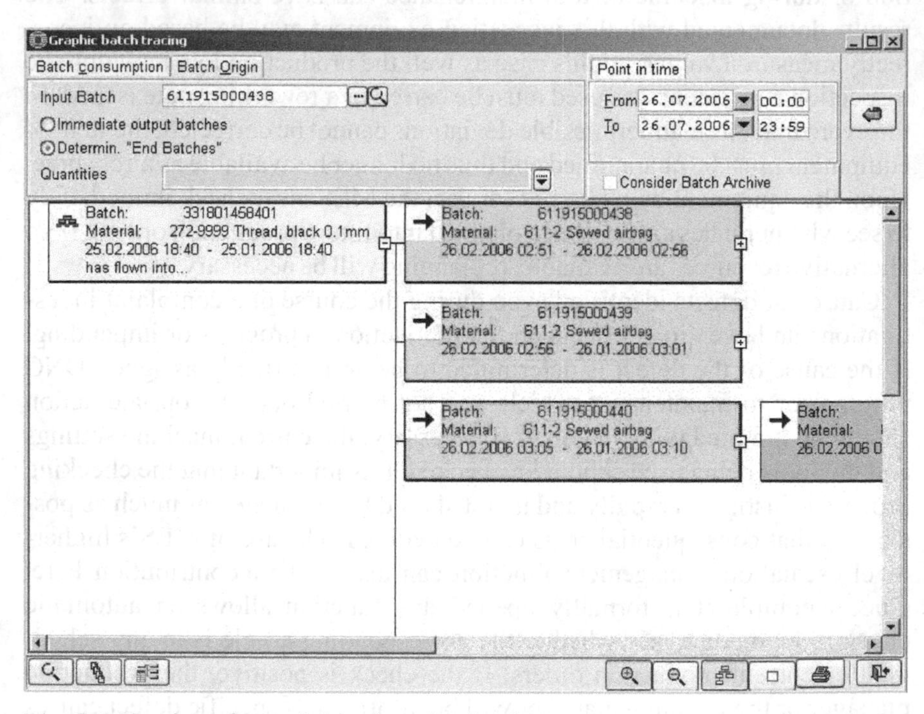

Fig. 8.6. Tracing of materials and production lots

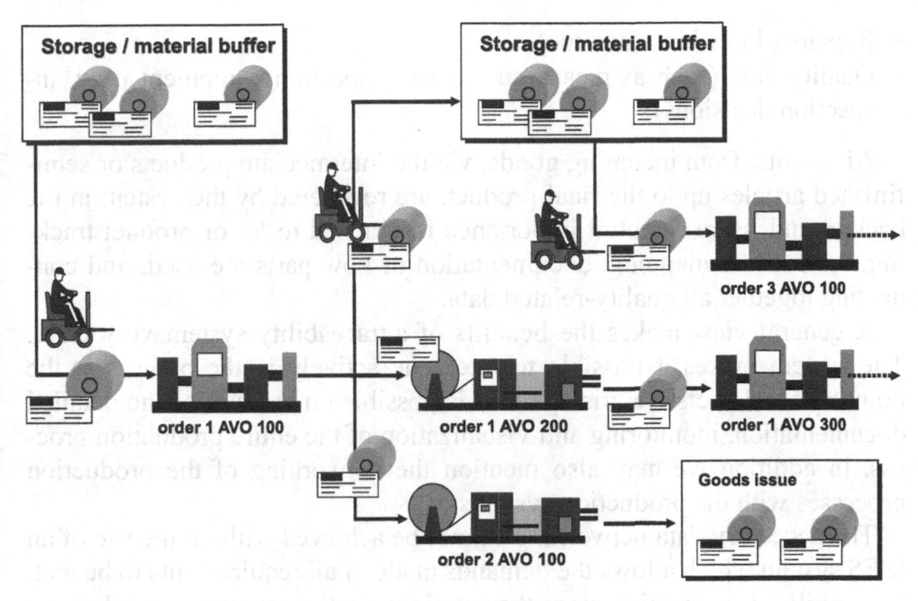

Fig. 8.7. Tracking of materials and production lots

Traceability is gaining in importance and not solely from the point of view of statutory requirements. It is also becoming increasingly important in the context of reducing production costs and thereby increasingly becoming a strategic factor on the corporate level.

In many companies quality assurance systems are established as island solutions. They cannot however provide complete traceability and in the event of a claim this can lead to problems. In connection with statutory regulations, the introduction of a traceability system is indispensable. Helping users to manage this requirement is a basic task of an MES system. To put it another way, MES is about ensuring the traceability of products.

All of the details of product creation are traceably documented in a traceability system. This includes information from all modules of an MES system, for example:

- Lots or batches coming in or being created
- Lot and batch attributes describing material (for example, weight, length, bond areas, date of manufacture, expiry date)
- Operating supplies used
- Machines used
- Process data obtained
- Individuals involved in the production process
- Tools used

- Repairs of machines and tools
- Quality data (such as measured values, inspection equipment used, inspection decisions)

All events, from incoming goods, via the intermediate products or semi-finished articles up to the final product, are registered by the system in the background. Here a central importance is attached to lot or product tracking, the process mapping, documentation of how parts are used, and connecting together all quality-related data.

A general view makes the benefits of a traceability system very clear. The system makes it possible to intervene actively in the process at the point where defects are arising. This is possible on account of the detailed documentation, monitoring and visualization of the entire production process. In addition we may also mention the networking of the production processes with the production systems.

The necessary data networking cannot be achieved without the use of an MES system which allows the demands made of all requirements to be met. Traceability is something more than straight quality assurance and delivers immediate economic advantages all along the value chain. Processes can be optimized and product lead times and inspection times reduced.

8.5 Analyzed and evaluated quality

Today it is more important than ever to obtain the information relevant to decision-making in a clearly prepared form. Evaluations and analyses in real time provide important information allowing well-founded decisions to be made or actions to be taken in real time. Appropriate data must be quickly available for every corporate level (from purchasing, sales, production and up to the topmost level of management).

In these days of global procurement of information, an isolated examination and analysis of quality data plays only a subordinate role. This does not mean that statistical key values and analyses of unadulterated quality data are no longer important. They will continue to be needed and indeed often form the basis for higher-level evaluations. However, achieving ever shorter product cycles, optimizing processes and being always one step ahead of the competition means that all potential means of evaluation must be exploited. It must be possible to react flexibly to rapidly changing general conditions. Here monolithic standard products with permanently defined integration points rapidly come up against their limits and the necessary expansions for their interfaces must be purchased at great cost, although nothing can be done about the absence of flexibility.

An integrated system offers the required flexibility, since by its basic structure it already allows access to all data from "quality-remote" areas, such as machine, tool, personnel and process data. Improvements in obtaining information within the company are the result and this forms the basis for process optimizations and the future success of the company.

8.5.1 Potential for improvement in production

Product quality is affected by many different factors. An exclusive analysis of quality data by means of

- Control charts,
- Statistical key values,
- Distribution tests,
- Main defects,
- Complaints,
- Inspection equipment calibrations, and so on

makes only a small contribution to improving the production process and the product quality. The potential for optimization is thus not fully exploited. What is of decisive importance is the inclusion of external influencing factors, which are also referred to as "quality-remote" parameters. These include:

- The acquisition and processing of process data
- Operating resources management
- Material management
- Personnel management

This is achieved with an acceptable level of effort and at low cost by the use of an MES system. While correlations within the product characteristics can be identified without a great deal of trouble in a CAQ system, such a system soon reaches its limits when process characteristics are included. It is precisely the higher-level analysis of product characteristics and process characteristics that has a considerable potential for improvement. The pressure or temperature in a machine can have a decisive impact on certain product characteristics. If a relationship of this kind is picked up by means of a higher-level correlation analysis it will be possible to achieve a sustained reduction in the scrap rate by monitoring the pressure or the temperature. It is conceivable that automatic monitoring will allow longer intervals between product characteristic inspections and this means savings in inspection costs.

This all-encompassing analysis of the machine data makes possible an additional optimization of the production process. By evaluating the reasons for idle times or the utilization ratio with respect to the articles and article groups being produced there, the best machine for the particular instance of production can be identified. A further improvement becomes possible when the machine-related reasons for malfunctions and for scrap are examined in relation to the product defects detected by inspections and the reasons for these defects. In addition to examination of the lead time there are more "quality-remote" factors which will allow optimization of value creation.

Another example of the advantages of a comprehensively networked data pool is a cavity-based and cavity-comparative analysis which takes into account the entire history of the mould. This will reveal the effects on product quality of repairs to or maintenance of individual cavities or whether individual cavities need frequent re-machining during repairs and maintenance.

One advantage of an MES system is that the data required for analyses are rapidly, systematically and inexpensively available from all departments. Since data acquisition and processing as a whole is based on uniform master data these evaluations can be regarded as secured. There is no duplicated management and maintenance of master data on the basis of, for example, reasons for scrap and malfunctions, types of defect, causes of defects. This means there is no risk of faults in quality management nor, as part of identifying the reasons for problems, in machine data collection being registered twice over or incompletely and being analyzed separately. Without this unitary approach to analysis it is difficult to define a clear structure for determining the number of reject parts.

8.5.2 Learning from complaints

An indispensable part of a comprehensive complaint analysis is that access is also possible to the tool data (maintenance and repairs) as well as to the machine evaluations (for example, reasons for malfunctions, scrap, machine idle time). It is important to clarify whether there were difficulties at the machine during the production time or whether a tool change took place on account of problems. The reasons for defects can thus be quickly identified and corrective measures taken to prevent them occurring in the future. As part of tracing there will be rapid access to all materials used, including the corresponding detailed information. This means that an answer can be obtained quickly to the question "on which machine, with which tool and by which shift or individual was the nonconforming product made". The consequence is that it takes less time to process complaints, which in turn is in the interests of customer orientation.

8.5.3 Six Sigma: putting a stop to waste

Despite the introduction of quality management systems, such as those based on ISO 9000:2000, QS-9000, VDA 6.1 or ISO/TS 16949:2002, there is still considerable room for improvement in processes and products. Experience shows that defect costs in companies run to as much as 30% of annual turnover. There is a reserve here which should be exploited. Further optimizations can be achieved by

- Reducing lead times,
- Reducing inventory,
- Increasing productivity,
- Improving delivery reliability.

Six Sigma is a suitable method for improving the quality of products and processes. It forms a basis for increasing customer satisfaction and for a sustained improvement in performance. The idea behind Six Sigma is to detect the number of faults in processes, to measure them and then to systematically eliminate them. Even if the core of Six Sigma is based on the application of statistical methods, the concept is considerably more comprehensive due, among other things, to the inclusion of findings relating to the processes. This is illustrated by the successful use of Six Sigma in administrative areas such as after-sales service and sales order processing.

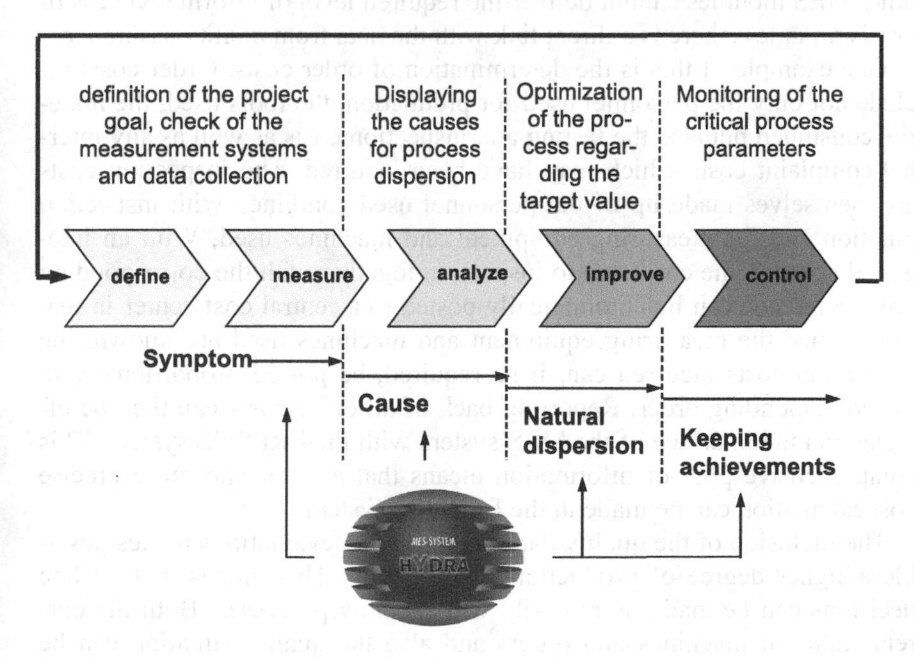

Fig. 8.8. The five Six Sigma project phases according to DMAIC

Restricting Six Sigma to quality management data will not suffice. Instead, information from all departments of a company, such as operating resources management, materials management and personnel management. Examples of this are the lead times for orders and the reasons for malfunctions and downtimes when a production stop occurs. This shows how important is the use of an MES system with its networked information.

The DMAIC cycle is frequently employed as a means of achieving process and product improvements. The DMAIC cycle has five phases and ensures that an improvement project is defined correctly and carried out using suitable methods.

Experience shows that when Six Sigma is used, waste in a company due to defects, overlong lead times and excessive costs can be drastically reduced within a few months. In addition, the results of Six Sigma projects can be applied directly in many MES functions such as resource planning and order planning.

8.5.4 Quality information: added value in the MES

As far as analyses are concerned an MES system has several advantages over monolithic standard systems. First, data from "quality-remote" modules can be used directly in quality analyses. In addition, many analyses in other MES modules cannot deliver the required level of informativeness or precision unless there is a direct link with the data from quality assurance.

One example of this is the determination of order costs. Order costs include not only the personnel used for production, the tools used, the material consumed but also the testing and inspection costs as well as any internal complaint costs which may have been incurred. The inspection costs are themselves made up of the personnel used combined with inspection duration and the measuring equipment and machines used. With an integrated solution the costs due to inspection together with the corresponding order reference can be automatically posted to a central cost center in real time. Since the measuring equipment and machines used are known, the calibration costs incurred can, if so required, be posted proportionally to the corresponding order. Reporting back of all order costs can then be effected via the interface of the MES system with the ERP/PPS system. This comprehensive pool of information means that a better and more precise cost calculation can be made in the ERP/PPS system.

The inclusion of the quality data in other MES evaluations makes possible a higher degree of production transparency. This means that ad hoc decisions can be made more easily with running processes. Both the current status of machines and orders and also the quality situation can be displayed by the MES in the form of a graphic machine park. The user is

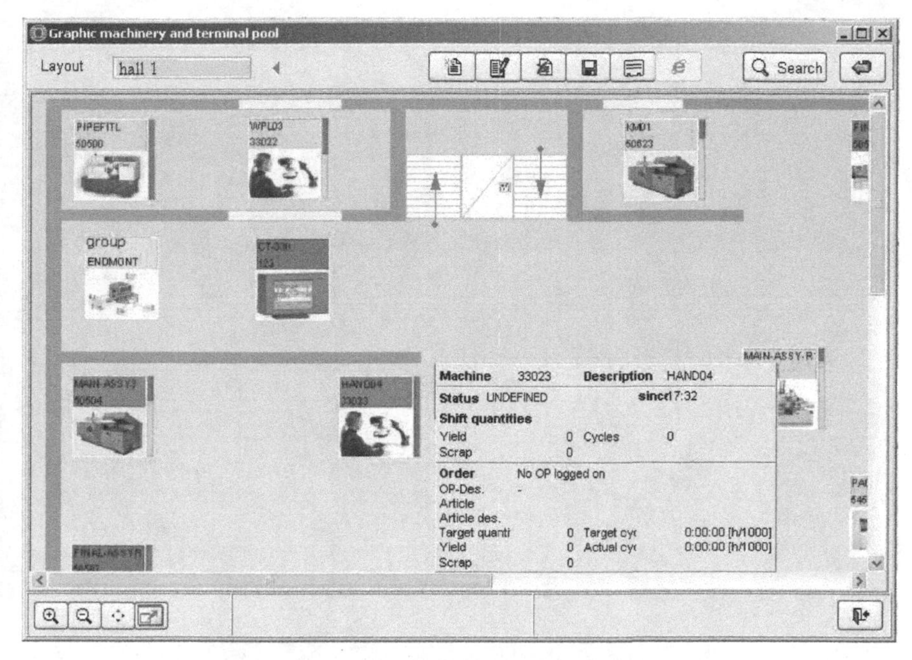

Fig. 8.9. Graphic machine park with quality information

aided in making decisions by being able to directly access detailed information in the form of control charts, quality non-conformances detected, action taken and ppm values.

In addition, actions which are important for quality assurance are in most cases also important beyond the circle of people using a CAQ system. By means of a system-wide action and escalation management function, an optimum distribution and processing of tasks is ensured as is the ascertaining of "real" effectiveness.

An MES makes it possible to compare at any time the current ppm value (number of defects per million possibilities) with the limit values prescribed in quality assurance. Via the MES the number of parts which have been produced is determined automatically and, taking into account the number of nonconforming pieces, the produced ppm rate can be calculated immediately. A warning can be given as soon as a limit value is violated and an escalation initiated. A necessary condition of this, however, is that the number of manufactured parts and nonconforming parts be registered centrally and not twice over – that is, in both quality assurance and production.

9 Personnel management with MES

9.1 Overview

Personnel is an important resource in a manufacturing company, if not the most important resource of all. One task of an MES system is to carry out effective and flexible planning of labor capacities for use in production. In a networked production department it is important to include not only installations and machines, orders and qualities in planning and optimization but also to a particular extent labor capacities.

The increasing importance of the "resource" of personnel in the production process has to do with two reasons primarily:

1. Since wage costs and non-wage labor costs are very high in industrial countries they have a great influence on production costs. As the result of the general trend towards globalization and the economic opening up of many countries in the Far East, workers in our latitudes are in direct competition with workers in other countries. To counteract this geographical disadvantage it is important to deploy personnel as effectively as possible.
2. The use of very sophisticated and highly specialized machines calls for a correspondingly high level of qualification on the part of the machine operator. It therefore becomes increasingly necessary to deploy employees on the basis of their capabilities and their knowledge.

Successfully overcoming the challenges which arise from this calls for effective solutions which reflect these requirements. The personnel management domain within an MES system provides tools which permit effective implementation of these tasks. A particularly important role is played here by the seamless integration of the personnel component into the overall MES solution thereby giving a complete and comprehensive view of the planning, the work flows and the results of production.

In addition to workforce planning, a company makes other requirements of personnel management in the fields of security, time registration and personnel management. Powerful MES systems also offer solutions for these points which enable the corresponding tasks to be mastered elegantly and at little expense.

9.2 Staff work time logging

An important function of personnel management in an MES system is staff work time logging, which in recent years has developed from managing the attendance and absence times of employees into a control system for personnel resources.

9.2.1 Tasks of staff work time logging

Staff work time logging is concerned with the three main aspects of time registration, time management and short-term manpower planning.

Time registration

Instead of the old time clocks, computer terminals are now used for registering the clocking-in/clocking-out times of the workforce. In addition, it is also possible to input breaks and the reasons for starting work late or finishing early. The terminal can also display up-to-date information about vacation days remaining, time credits or recent clocking-in or clocking-out times. Messages can also be sent for employees to read at the terminal and these messages can, for example, be displayed when the employee clocks in.

Time management

The task of time management is to ascertain the actual time worked by rounding off the clocking-in and clocking-out times and debiting the work breaks. Following comparison with the required work time laid down in the work time model, it calculates the overhours or underhours, should this apply. These can be paid out or collected in a time account. In addition to time accounts, the vacation account is also maintained. The working hours worked by the employee are posted by time management to pay types on the basis of configurable payment rules and these are added up at the end of the month and transferred to the payroll department. Here we understand work time model in a broad sense which ranges from the familiar day/shift time, via flextime to monthly, annual or even lifetime working hours. With workflow management, procedures such as applying for vacation or correcting incorrect clocking-in or clocking-out events can be reproduced without the use of paper. In addition, current overviews display present and absent employees, trends in the hours worked and statistics for times of absence.

Attendance/absence overview

Table | Board

Person ... to ...

Comp. Area Cost Center

Persons Present ☑with Order No. ☑without Order No.

Persons Absent ☑planned ☑unplanned ☑free

Please drag a column heading here to group by this column

Person			Status				Absence		
Comp.	Person	Name	Status	Date	Time	Present at	Absence	Date from	Date to
BSP	667	Smith, Christian	Absent since		6:03				
BSP	885	Burger, Simone	Present since		7:01				
BSP	50014	Albert, Claudia	Planned absent			21.03.2006	Leave	01.03.2006	19.03.2006
BSP	50201	Meyers, Harald	Free since		13:00				
BSP	59881	Brown, Albert	Free until		15:00				
BSP	85741	Meyers, Joseph	Present since		5:57				
BSP	93856	Miller, Steven	Present since		8:23				
BSP	96665	Hanson, Betty	Unplanned Absence since		9:00				
0001	10002541	Berry, Claire	Free until		23:00				
0001	10002630	White, Susan	Planned absent			08.03.2006	Illness	24.02.2006	07.03.2006
0001	10002854	Baker, Jake	Present since		7:55				
0001	10003645	Hoffman, Paul	Free						
	12								

Fig. 9.1. Overview of attendance and absence

Short-term manpower planning

Short-term manpower planning is concerned with which employee works at what times and on which shifts. Here the occupational activity and qualifications of the individual employees play a major role. Once adequate manpower for a shift has been secured, planning can go on in the next step to determine at which work center or on which order the employees are to work. Even when orders are planned in at the control station, short-term manpower planning can be used to check whether the current order situation is covered by sufficient personnel with the corresponding abilities.

9.2.2 Time management in the MES or ERP system

While time registration is a fixed component of an MES system, time management can often be carried out even in the ERP or payroll accounting system. The advantage of this approach is that an interface with the payroll department is not required and integration into the general accounting and controlling departments is a simple matter. On the other hand the advantages of using time management within the MES system include:

- The attendance time determined can be compared with the work order times reported in the PDA. This comparison is important for ensuring that all of the hours worked by the employee have been recorded in the PDA since these data are required for the controlling department or for incentive payments.
- Running time management within the MES system means that periods of absence can be registered, incorrect clocking-in and clocking-out events corrected and overtime approved on a decentralized basis by the foreman using the interface familiar to him from, for example, PDA. There is no need to install an additional workstation.
- Short-term manpower planning requires that the scheduled work times of the employees be held in the MES system. This makes it possible, for example, to take the resource of personnel into account when planning orders at the control station.
- While payroll systems have a standardized interface for the transfer of salary types, interfaces for inputting work hours performed from the time management system to allow comparison with the PDA and with scheduled work times or absence times for the purpose of short-term manpower planning will in most cases need to be implemented on a project-specific basis.

One example of an ERP system which includes a time management function is SAP-HR. When using SAP a decision must be made as to whether the MES system should be used as a subsystem for registering clocking-in and clocking-out events or whether the time management function in the MES system should be used. In either case the HR-PDC interface is used.

9.2.3 Flexibilizing work hours

Trends like "just in time" are concerned with reducing the costs of stock keeping and of current assets by always manufacturing exactly what the customer wants at that precise time. A consequence of this is considerable fluctuations in capacity loading in production and in many management departments.

In addition, seasonal variations in the demand for particular products result in a different distribution of labor requirements. At the same time, however, personnel costs take up a high proportion of production costs which means that preventing unproductive wages is an increasingly important task of personnel planning. These and many more reasons besides make an increasing flexibilization of working hours necessary.

One simple way of satisfying this requirement is to introduce a time account. Irrespective of whether the employees themselves determine the movements of their time account on the basis of the amount of work available or whether building up or cutting down the hours in the account is controlled by the company, a time account provides a way of matching the labor deployed to the current demand.

In line with this, different types of accounts are used. If the employee himself is responsible for running his account, this type of account is called a flextime account. For their part, companies use the flextime account for controlling additional work and shortages of work. In the case of a seasonally fluctuating inflow of orders, an annual account can be used for ensuring that the work time evens out over the course of the year. A high demand for labor sustained over a relatively long period can be covered in a lifetime working hours account. With this type of account, the additional hours worked can give the employee the opportunity of retiring earlier.

Any overtime which is required can be worked before or after regular working hours. In the case of companies which are already working three shifts so as to achieve a high level of machine utilization, additional shifts can be planned in at weekends.

It is, however, still possible to use flextime with three-shift operation. What matters to production is not that the employees start and finish their work at particular times but that enough personnel are present to operate

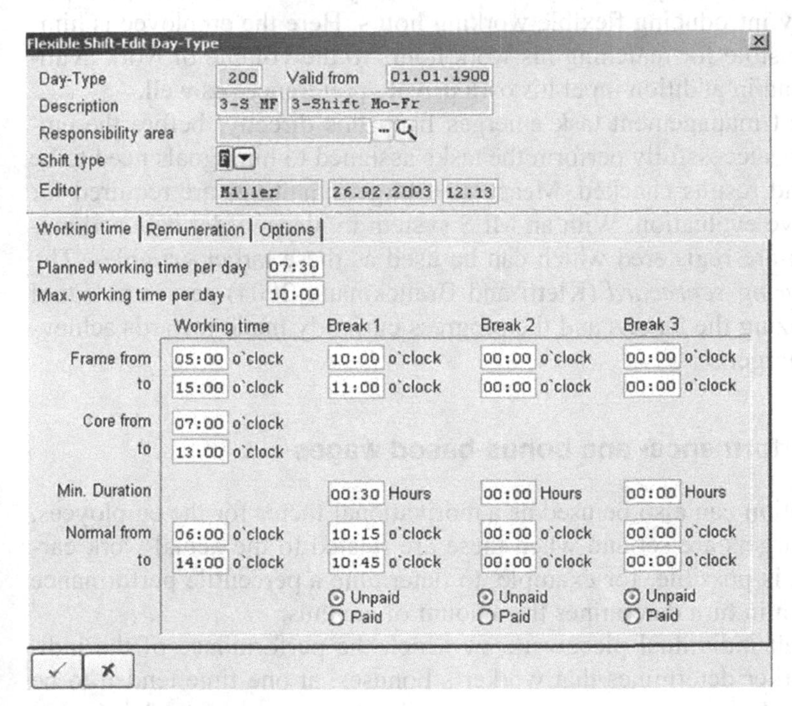

Fig. 9.2. Flexible shift-day type

the machines. When there is a flexible shift handover (which is agreed by consultation between the employees involved), both the requirements of the company and also the wishes of the employees can be satisfied.

9.3 Motivation and personnel management

A study conducted by the Gallup Organization came to the conclusion that in Germany in 2002 only 15% of employees were motivated in their work. 67% of those in employment went to work without any motivation and 18% actually worked deliberately destructively. Comparisons with previous years revealed that the number of motivated workers was falling. This is shocking when you are aware of what potential is available in motivated workers and when you imagine what would happen if some of this power were even directed against the company.

The question therefore arises automatically as to how the motivation of the employees can be increased. However, many managers are not aware that it is their job to release this potential. Management seminars are required on all management levels to clarify what this task means.

Various motivators exist for employees. One way of improving motivation is to transfer responsibility to one's employees. This can, for example, be done by introducing flexible working hours. Here the employee is himself responsible for matching his work hours to the volume of work available and can in addition input his own private preferences as well.

The next management task emerges from this directly: before the employee can successfully perform the tasks assigned to him, goals need to be defined and results checked. Measurable target quantities are required for an objective evaluation. With an MES system in place, order data and machine data are registered which can be used as these target variables. The *manufacturing scorecard* (Kletti and Brauckmann 2004) serves as a tool for visualizing the targets and the progress currently made towards achieving those targets.

9.3.1 Performance- and bonus-based wages

Remuneration can also be used as a motivational factor for the employees. Specific targets are set and when these are related to the actual work carried out it is possible, for example, to determine a percentile performance level which in turn determines the amount of a bonus.

Although individual piecework, in which the performance of the individual worker determines that worker's bonuses, at one time tended to be the standard approach, nowadays group bonuses are preferred in many

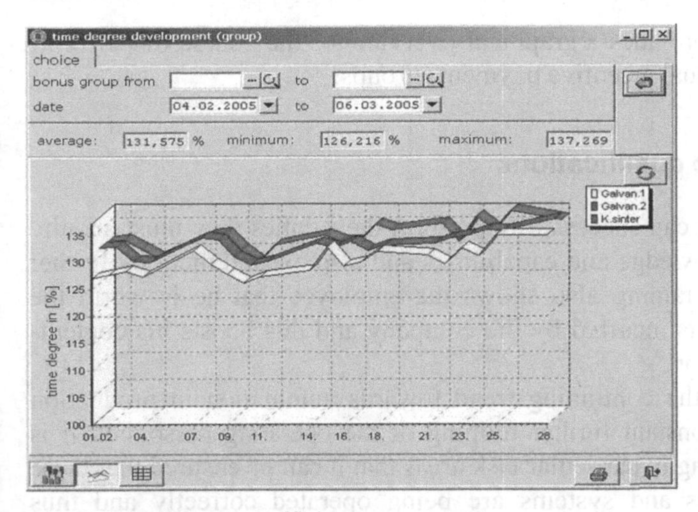

Fig. 9.3. Performance efficiency curves

companies. One advantage of group-type performance-based remuneration is that this approach encourages the employees to work together. In addition it is possible to include in the bonus even those employees such as charge-hands or lift-truck drivers who are only indirectly involved in the production process.

It is primarily the order notes which form the data foundation for determining the performance payments. However, in an MES system, machine data, data from staff work time logging, or quality data from the CAQ module can also be input into the performance wage without the need for interfaces. In comparison with a performance bonus, in which a comparison is made of target and current times, a utilization bonus also takes statistics into account, such as the number of OK pieces and the number of scrap. Utilization bonuses thus make sense when the actual time is primarily determined by the machine work cycle and the worker has virtually no influence on it.

Another example of integration of performance-based wage calculation for the purpose of time registration is the comparison list with which a check can be made to see whether the total attendance time of the employees has been posted to the orders and whether the underlying data are complete for calculating the performance wages.

Since there is great variation in collective agreements in the various sectors of industry as well as regionally and since there are great differences in how performance wages are calculated in individual companies, the performance-based wage calculation function requires a great deal of flexibility and needs to be readily adaptable to payment rules. The performance efficiency curve provides a graphical overview of the course of performance in the individual incentive payments groups.

9.3.2 Employee qualifications

Before employees can successfully perform their duties they must acquire the necessary knowledge and capabilities and also expand them by further training. Further training also shows the employee that he is worth the corresponding costs incurred by the company and this boosts his contentment and motivation.

It is precisely the continuing trend towards automation in production which calls for constant further training of the operating personnel. It is only by specializing in particular task areas that it can be ensured that these complex machines and systems are being operated correctly and thus reaching a high level of productivity.

What is important to the company is that its investment in its employees is implemented in a way which yields a profit. For this reason one of the tasks of the MES system is to register employee qualifications and display them close to the production line. For example, a plausibility check is carried out when an order is opened to ensure that the person handling the order has the necessary qualifications. This procedure will improve the quality expected in the finished products and also increase the motivation of the employees since they are not be used for tasks for which they lack the requisite knowledge or skills.

9.4 Short-term manpower planning

Current studies show that around 60–70% of all companies use a standard spreadsheet program as a tool for planning the deployment of personnel. Allocation of which employees will work on the scheduled orders at which work centers is also frequently carried out using wall charts.

The problem with these methods is that data such as the vacation applications of employees have to be maintained in several locations. Redundant data storage of this kind means unnecessary extra expenses and results in individual data statuses drifting apart. This in turn is the cause of errors in personnel planning and the economic efficiency of production drops due to a lack of personnel or to an oversupply of personnel. Although the data are maintained in several locations they are often not available in other important locations. For example, when planning in the orders it would make sense to be able to compare the planning against the labor capacities available.

Successful presentation of the requirements relating to planning personnel deployment calls for several tools which can display labor capacities from different points of view and interdependencies between them.

9.4.1 Vacation and shift planning

To be able to process vacation applications from his staff the supervisor must have an overview showing whether enough personnel are still available for a controlled process flow. In this regard it is not sufficient to take into account the number of employees who are scheduled to be available but the jobs they perform and their time account balances must also be considered.

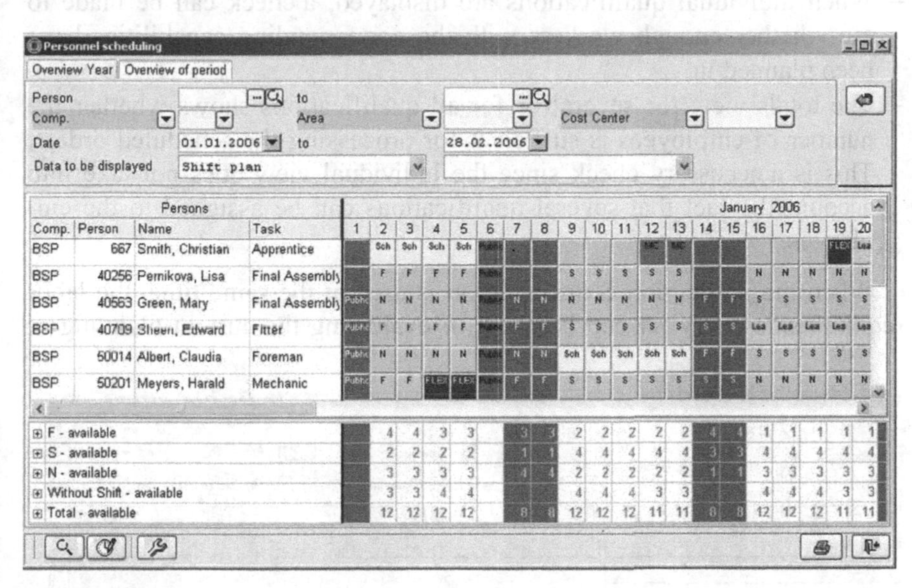

Fig. 9.4. Vacation and shift planning

While vacation planning is often carried out a long time in advance, shift planning tends to be at medium or short range. The task of shift planning is to ensure that the individual shifts are allocated sufficient personnel. Since this is based solely on data from staff work time logging, empirical values or target values from the ERP system will be used if necessary.

A vacation and shift planning module is used primarily by foremen and shift foremen. For this reason this module will have to be operated intuitively and simple planning functions provided. The result of this planning is the shift plan, which is made available to employees as a printout or in electronic form or which the worker can call up at the terminal.

9.4.2 Checking labor capacities during detailed planning

The labor requirements overview shows employee requirements grouped by qualifications. The actual requirements either derive from the orders scheduled for a work center or can be defined permanently on the basis of the work center. Opposite them, the screen shows the scheduled labor capacities together with the qualifications assigned to them. This results in a display which shows coverage of labor requirements, deficiencies in personnel, and excess coverage of requirements.

Since an employee may have several different qualifications, the labor requirements overview is used under different aspects:

- When individual qualifications are displayed, a check can be made to see whether enough workers with the corresponding capabilities have been planned in.
- The totals view for several or for all qualifications shows whether the number of employees is sufficient for processing the scheduled orders. This is a necessary check since the individual view does not take into account the fact that several qualifications can be assigned to the employees.

When the production control station is used at the same time, the labor requirements overview can be used for evaluating the current planning or

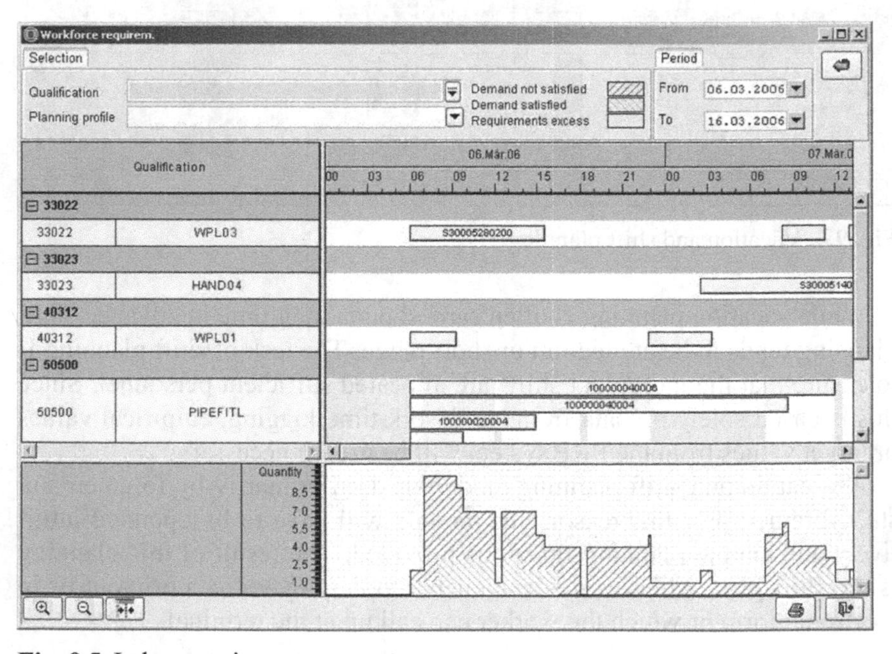

Fig. 9.5. Labor requirements

a planning simulation on the basis of the labor capacities. This shows the advantage of an MES system in which individual modules can be mutually integrated without interfaces needing to be defined or implemented.

9.4.3 Allocating employees to work centers

The personnel assignment module is used for assigning employees to the work centers. In a similar way to the labor requirements overview it determines what workers with particular qualifications are needed on the basis of the orders or this is defined directly at the work center.

The bottom part of the personnel assignment screen shows what employees are available while the top part shows the work centers to be filled. By selecting a planning profile the view can be cut back to a single planning unit with particular work centers and employees. When personnel are assigned manually to the work centers, a plausibility check is carried out to see whether the worker in question has the necessary qualifications.

Employees can also be planned in automatically. Here the workers are assigned automatically to the work centers on the basis of their qualifications. If several workers have the same qualification, it can be determined with the aid of a ranking scale which employee has the best command of

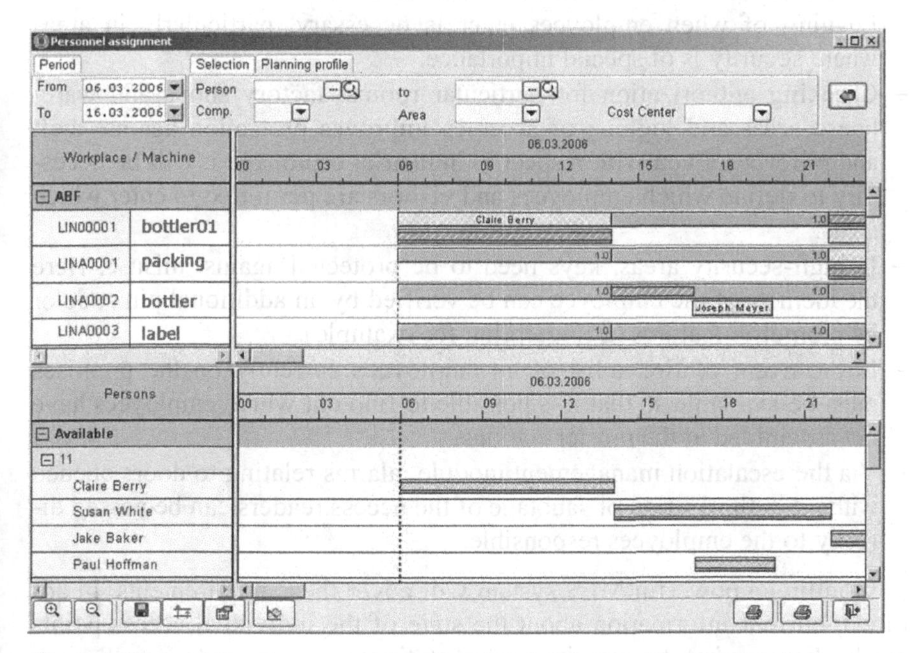

Fig. 9.6. Personnel assignment to resources and orders or operations

the capabilities required. With these data, the personnel assignment function attempts to identify the best plan possible whereby the work centers are completely filled with the best qualified workers. The result of personnel assignment is a personnel deployment plan which is printed out for the employees or which can be viewed at the terminal via an information key.

9.5 Security in the manufacturing company

The terrorist attacks of September 11, 2001, on New York and Washington changed the world. One consequence of these events is that many companies need a higher level of security. But there are yet other reasons for monitoring the entrances to and exits from company premises and also the doors between different rooms inside company buildings:

- Unlike conventional locks, with an access control system (ACS) time-based authorizations can be assigned. This means that it is possible, for example, to ensure that certain employees will only be allowed access during working hours on weekdays while other employees will have 24-hour access to their place of work, and even at weekends as well.
- With a conventional lock, a lost key often means that the entire lock has to be replaced while with an access control system the lost ID card loses its authorizations as soon as a new card is issued.
- Logging of when employees enter is necessary, particularly in areas where security is of special importance.
- Checking authorization for particular rooms, factory shops and warehouse areas and logging of accesses improves protection against theft and pilfering. Even with respect to industrial espionage it will be necessary to define which employees and visitors are permitted to enter which areas.
- In high-security areas, keys need to be protected against misuse. Here the identity of the employee can be verified by an additional pin code or by biometric features (a fingerprint, for example).
- In the event of fire, a list of all employees currently on the premises must be available so that it is possible to find out which employees have not assembled at the muster stations.
- Via the escalation management module, alarms relating to doors opened without authorization or sabotage of the access readers can be passed directly to the employees responsible.

Installing a powerful MES system will cover these requirements. In addition, current information about the state of the individual access points can be displayed at the security control station:

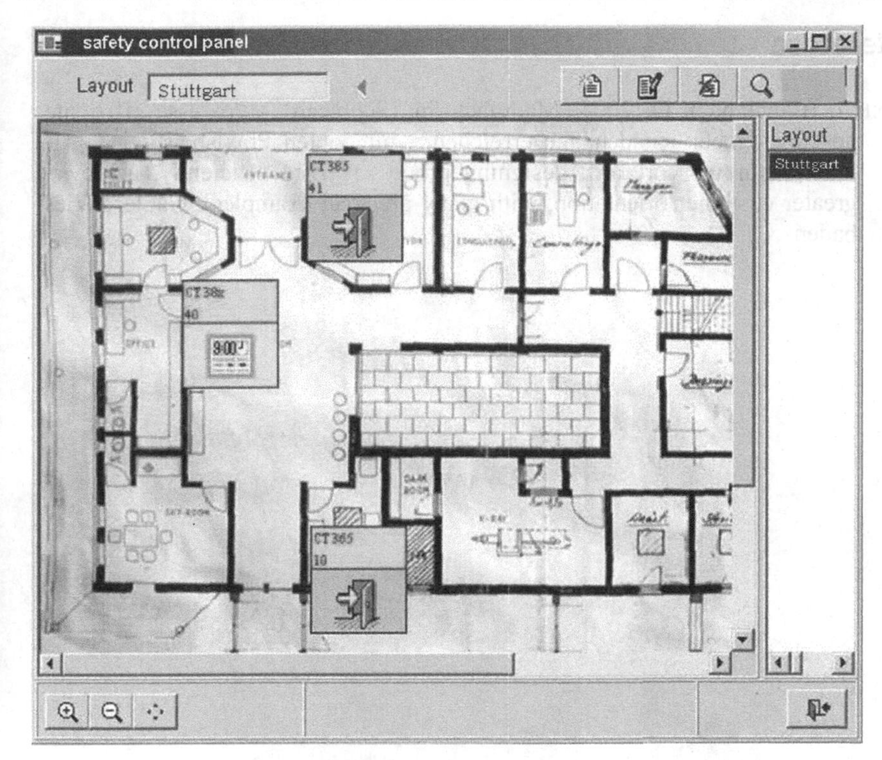

Fig. 9.7. Security control station

9.6 Outlook

In 2004 the term "human capital" was picked by an independent panel as non-expression of the year. Irrespective of whether this was a good or bad choice, the expression does show the increasing importance of employees to the company.

After many years in which financial criteria were primarily used for evaluating a company, it is a new approach to include the employees with their knowledge and their capabilities in the analysis. It necessarily follows from the great importance of personnel that the personnel must also be included in MES system as an important component.

Here it is especially important to regard tasks such as staff work time logging, short-term manpower planning and access control not as individual disciplines but rather in conjunction with other modules, such as detailed planning, for example. Accordingly the strength of an MES system is that production can be considered from a large number of different angles and thus a comprehensive overall image of production be obtained.

Literature

Kletti J, Brauckmann O (2004) Manufacturing Scorecard – Prozesse effizienter gestalten, mehr Kundennähe erreichen – mit vielen Praxisbeispielen [The manufacturing scorecard: designing processes more efficiently, achieving greater customer orientation, with many practical examples], Gabler, Wiesbaden

10 MES with SAP

10.1 Motives

Due to its great popularity and high degree of standardization SAP has in two regards a special importance among ERP systems on the market. In the first place, developments for an MES vendor in the SAP environment pay off due to the high level of market penetration; Second, SAP is one of the few suppliers in the ERP environment who actively advocates an independent MES system outside mySAP ERP in order to be able to offer their customers vertical integration. It is precisely in the manufacturing environment that a partner concept for so-called manufacturing applications has been massively beefed up since mid-2004: the "adaptive manufacturing" partnership leads all other efforts.

With this initiative and the corresponding publications SAP is clearly demonstrating that for many industries and areas of application the use of an independent MES is to be recommended to allow exploitation of the full functionality of the SAP ERP and logistics solutions. On the SAP side this depends on integration of the MES solutions of the partners.

The idea of integrating an MES system into the ERP world is gaining a new importance at SAP and demands of an MES – and thus, of course, of an MES vendor – a whole series of interfaces and integration technologies. It is, however, of decisive importance to the SAP user – in other words, the manufacturing company using mySAP – that he does not lose sight of what is really necessary to secure the quality of the application. Here the suitable technology itself should only form the basis of being able to build an application. SAP offers a technology platform for implementing a "wide" range of applications with the MES very deeply integrated into the SAP world. SAP puts its SAP NetWeaver technology platform at the center of these efforts. Accordingly, labels such as "Certified for SAP NetWeaver" initially mean only that the vendor in question is using the technology but not that any added value has been generated here in the application itself.

The certificate "Powered by NetWeaver" aims at considerably more in the direction of application. It confirms that a partner has integrated his application on the NetWeaver platform into the SAP applications.

It is very difficult for the user to separate technology and application, especially in the SAP environment. Lavish marketing campaigns on the part of SAP and the SAP environment do not always make a contribution here to increasing transparency. The technical terms and buzz words alone increasingly compel the medium-sized manufacturing company to ask the question: "what is the benefit for my company?". The answer to this question is the motivation behind what follows. We also look at the definitions thus provided which relate to integration of an MES system into mySAP ERP.

10.2 Integration of the MES into the SAP environment

10.2.1 Development of the MES in SAP history

In the SAP environment in particular, the beginning of the 1990s saw a simple feedback system, which was nothing more than a straightahead data collector, being offered on the market as a production data acquisition application (PDA) or subsequently even as an MES system. Frequently these systems consisted or consist of a more or less complex feedback device combined with very simple interfacing software for SAP R/3. Following data acquisition, the systems "deliver" the data to the R/3 system via conventional data interfaces. Once systems of this kind are in service, their structural shortcomings become evident. Due to their structure, these systems are not capable of preprocessing data or even carrying out more comprehensive plausibility checks. Via the acquisition mechanisms it was in fact possible to replace only the manual records, the so-called workshop recorder, with an electronic data recording facility. The lack of preprocessing or plausibility checks for the data resulted in an unsatisfactory quality of the actual data in the SAP system. The data had to be reprocessed to improve the data quality. The absence of objectivity in the data which were obtained in this way automatically gives rise to doubts as to the correctness of operational parameters and control inputs for production "calculated" in R/2 on the basis of data of this kind. Any attempt to get better data quality while retaining the same system architecture will be bound up with high costs for the manual maintenance of the data and for the IT infrastructure required to obtain better system performance. Furthermore, the result of this attempt was that the data acquisition/PDA system and the SAP applications above it – basically, merchandise management (MM) and production planning (PP) – intermeshed so tightly that even minor changes in the production process involved extensive modification of the software in the data collection system, at the interface and in R/3. This period also saw the emergence of approaches in which set-point and actual value status messages were generated so as to avoid IT costs. Of course this meant that no improvement in the planning data could be expected, nor even realistic statistical cost accounting. For rigid production processes which have reached a steady state, a manufacturing company can under certain circumstances dispense with continuously updated status messages about the quality of the production process. In an age of increasing customer-driven flexibility this is no longer acceptable in a modern production line.

10.2.2 Requirements for an MES in the SAP system environment

The consequence of this was that subordinated systems had to be given a smarter design, in other words, brought closer to the process, with the aim of improving data quality by means of plausibility checks close to the process.

What close to the process means in this context is carrying out inspections reactively on the basis of the current production situation and the control inputs from ERP (target dates, target quantities, inspection and test plans) and reporting the results to ERP. These results trigger actions there or, by supplying the user with the corresponding information, put him in a position to be able to intervene in control.

The technical aim as regards IT is naturally to check the flood of data in the mySAP ERP system by preprocessing it and thereby reduce the IT costs for the SAP system. Examples of this include reporting milestones in materials management (MM) while transportation of material together with all material movements is secured in the MES, or accumulating quantities and times and time-ticket-related status messages in SAP PP.

At the same time it is important that mySAP ERP and the lower-level MES be adjusted to each other in such a way that no redundancy results. In other words, the systems must be so "meshable" with each other that for the user they represent a single across-the-board solution and the user, for example, only needs to carry out maintenance of master date in one system. The functions of the two systems should therefore supplement each other and under no circumstances overlap. This is the demand for a dedicated implementation in a manufacturing company.

A widely branched manufacturing company – in other words, a company which although it has a central IT department also has several production locations which come under different sectors of industry – will have another important requirement. The MES should be flexible enough to be configurable to the different production organizations but still ensure uniform integration with the central SAP system.

Depending on the production organization or sector of industry, one production location may need a shop-floor control function in the MES system while another location does not need this function or it is available in mySAP ERP.

10.2.3 Representation of levels in a manufacturing company

On the basis of this question, it is important first of all to structure the processes and system functions which are found in a manufacturing company and to ascertain the system (SAP or MES) to which they can or must be assigned.

In some very recent publications, SAP has defined the tasks and functions of a manufacturing company under the aspect of their assignment under the term MES. From these publications a levels concept can be derived which presents the processes and functions. These functions are grouped into levels.

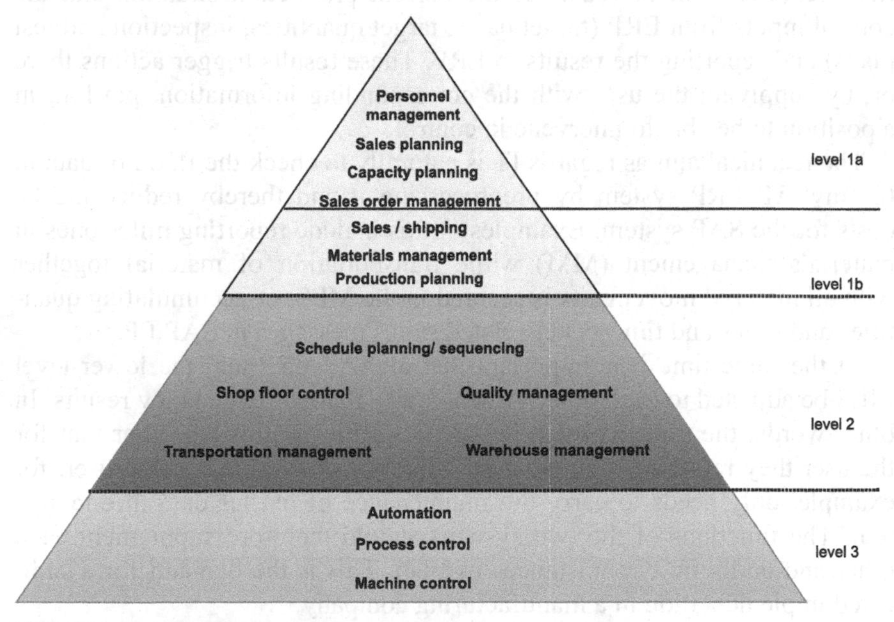

Fig. 10.1. Functional levels in a manufacturing company

Level 1a: Corporate planning

On the topmost level are found the classic ERP functions such as sales planning, capacity requirements planning, customer order management, sales, and shipping.

Level 1b: Planning and material planning for production

Level 1b includes the functions for product planning, production flow planning and materials management. This also includes the supply chain management function which controls all logistical processing on corporate and subsidiary levels as well as logistics between customers and suppliers. This is to be distinguished from transportation management, or so-called intra-logistics (technical logistics), in other words, material control within the production department of the company, which is described below. On this level too we also find the SAP applications of APO. Here the time horizon for planning work is in all cases medium or long range.

Level 2: Production management

On Level 2 are found all of the processes which are largely used for implementation of the production plans produced on the level above. This is where flow of material and production itself are controlled.

On this level too are found functions and tasks such as reactive planning – in other words, production control itself, also referred to as shop floor control. The time horizon here is short range.

Level 3: Automation

Depending on the industry and/or production structure, this level includes functions which in the process industry/continuous production are largely handled by process control systems. In discrete manufacturing these functions for controlling the production process are for the most part mapped on the level of production control (shop floor control). On this level we also find not only all of the functions and processes in the manufacturing company which control the machines and installations but also those which are responsible for the exchange of information and control parameters to and from the machine or machine group.

10.2.4 Corporate processes in mySAP ERP and the MES system

In what follows, these functions and processes are assigned to the IT system mySAP ERP and to the MES system. For those levels situated "outside", this assignment is conspicuous. The functions on Level 1 corporate planning are the classic disciplines of the ERP system and are covered by mySAP ERP. Even those functions for production planning and for the medium-range planning of materials and resources are for the most part implemented consistently in mySAP ERP.

The functions on Levels 4 and 5 are mostly to be assigned to the MES system or should at least be capable of integration into an MES.

Control of the production process is carried out in the MES system (process control function). This includes the classic data acquisition functions and naturally also the integration of the vital technologies for connecting machines, machine groups and process installations. The fundamental requirement is that all of the data are made available for the higher levels. This is the basis for ensuring the requirement for so-called vertical integration is satisfied.

However, since it is more and more often assumed in the manufacturing company that processes do not necessarily run in their entirety within a single system but rather need to be supported vertically over several system levels, in addition to the functions being assigned to one system a "considerably more interesting" part of system integration is the "consolidation" of functions from several systems into a single process mapping.

From the technical aspect of IT, this area of "overlapping functions" largely includes the functions of Level 3 of the present model. Depending on the production organization or sector of industry these processes can be entirely implemented either in MES systems or in SAP systems. As an alternative, however, and this does appear to be how things will be in the future, they can be broken down into sub-processes and mapped into both the SAP and the MES systems.

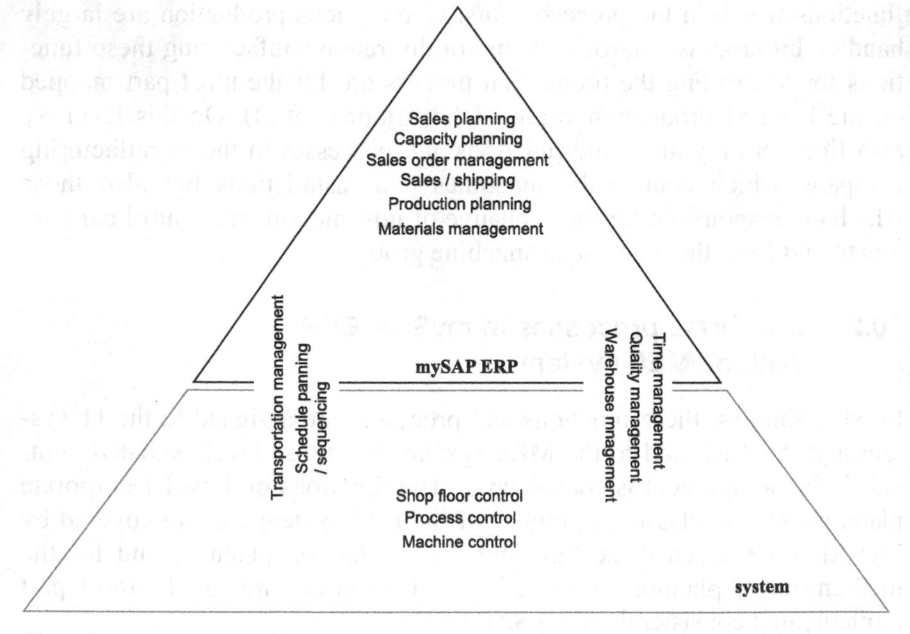

Fig. 10.2. Integration of manufacturing company processes into the IT landscape

Relevant processes in the MES system

As has already been mentioned, there are functions or processes which are mapped in the MES either completely or even only partially. The connection between the MES and mySAP ERP must be so flexible that the following functions, depending on the application and the production organization, can be "switched on and off" in a company.

Scheduling and sequencing, production control

While scheduling and sequencing in the sense of production control is, in the case of quantity production, for example, carried out on the medium to long-term basis – in other words, days in advance on the production plan – there are some sectors of industry and production organizations in which such plans are drawn up only just before production begins or which have to be modified while production is in progress. The term "reactive planning" was coined to differentiate between these planning methods. This term means the creation of an executable production plan which was optimized with respect to production conditions at the time of its creation. Reactive planning is necessary when, for example, different machine groups and production equipment suffer from technical restrictions on account of the materials being used. If the problematic properties of the materials are not identified until production is actually under way, it will be necessary to change the planning directly at the machine group or to do the actual planning work or production control there.

One example of this is the steel industry. Due to the properties of the material, planning changes are decided immediately before or during the production process, something that can result in over- or undersupply for the production order.

If this also affects other production stages which require the higher-level planning systems such as the SAP modules PP, MM or APO, they will have to be controlled via the systems affected. In this case it is the job of the MES to supply the SAP modules with information synchronously or to call up new planning requirements or even a new plan, or to have a new plan prepared.

Transportation management

Depending on the manufacturing sequence and infrastructure of an individual manufacturing company, the term transportation management has some quite different meanings. Transportation management definitely does not include the transportation of materials from one production plant to another, to a distribution center or directly to the final customer. Normally these functions are handled within the SAP system by supply chain management. In certain production industries, such as the steel industry, there are also great distances to be bridged as well. Between individual production areas such as the blast furnace and the hot-rolling mill, transportation logistics within the plant are an important and completely independent task for supply chain management. The same also applies however to manufacturing companies which on account of their growth history have very difficult infrastructural conditions (parts of the plant separated by public roads, parts of the plant at great distances from each other although at the same location).

Here transportation management and transportation logistics belong partially or completely to the production process and must be handled with a similar flexibility as the production process itself. It makes sense in this case to map transportation management as an independent logistics function within the MES system. The same is also true of the typical materials arising during production (WIP = work in process). The intermediate materials thus created have only a "short life" and for this reason it is not necessary to manage them in their own planning stage within materials management (MM).

In addition to registering and tracking intermediate materials in the production process, a decisive role is played on the transportation management level by the documentation of the creation process for customer end-products. In the pharmaceutical, food and automotive industries, lot and batch tracking down to the tracking of individual parts is becoming more and more important. In the fields of tracking and tracing, SAP is providing more and more functions in its industry-specific solutions, functions which must however be integrated by the MES into the production process.

Quality management

It is increasingly true of all industries that quality management is not an autonomous process in the manufacturing company but rather that it has to be integrated into the production process itself: this is the so-called in-process quality assurance. Employees in production itself should check the quality of the product, correct settings which affect the quality of the material or of the final product, or simply collect quality data. The typical expression for this is the worker self-check.

Support for this process requires that the data acquisition functions at the machine or machine group also include the monitoring functions for quality assurance. Control of this quality management function and archiving of the data does, however, come within the SAP QM. Whether acquired data or events are relevant to quality assurance or to the product documentation must be determined directly within the production process.

The production process facilitates everything required for in-process product documentation. Transportation management and quality management run on a single level within the MES system. At the time when the data originated, links between product and quality data and also measured values are created here on line.

Time management in production

In a company, time management up to and including calculation of wages is a constituent part of the personnel management system and thus a part of SAP HR.

It is possible to depart from this "rule" when performance-based wages are to be calculated in a manufacturing company for the productive employees in production.

In this case it is important that data are preprocessed immediately the data arise and are secured for data maintenance at a time close to the workflow. For example, if a foreman looks after personnel entries actually during the current shift rather than a time administrator not doing so until the next shift or even the next day, it will be possible to make a real-time comparison between the personnel and quantities signals and also check their plausibility. This means data quality and security for the production process and also an appropriate and above all effective employee remuneration conforming to the product quality and productivity.

For companies in which performance-based remuneration plays no role in production, it makes sense to map time management in the classic manner within the SAP environment (HR module). This thus reduces functionality in the MES to the simple acquisition of time events such as "clock in", "clock out", "break" or similar. But the MES will also have to satisfy requirements relating to plausibility checks on employees in production. In the interests of an overall solution it is also important that the system (SAP plus MES) supports both alternatives.

10.3 MES as an integrated solution in the SAP system

The levels concept, which was described in the previous section, and above all the overlapping of functions and processes which is envisaged, calls for a flexible and scalable method of linking the SAP system to the MES system.

For this purpose SAP provides, via conventional interface technology (in other words, the simple exchange of data between two systems), a large number of methods and procedures which offer a wide range of possibilities for linking up a lower-level MES system.

Since mid-2004 with the release of its product SAP NetWeaver, SAP has been pursuing the goal of bringing together all existing integration procedures in this product. SAP NetWeaver thus automatically plays a central role in the integration of an MES system in mySAP ERP.

10.3.1 Importance of SAP NetWeaver for integration of the MES

The aim is to implement NetWeaver as a kind of integrator for all applications in a company. It should be possible to swap round applications or sub-applications as desired. This also means that mySAP ERP must automatically break back down into individual applications (so-called services) if it is to have the necessary flexibility as an application. According to an SAP road map, it is not until 2007 that mySAP ERP will have full NetWeaver capability.

Since the NetWeaver concept satisfies the ESA* definitions, taken consistently this also means that that the SAP applications themselves are treated on an equal footing with the company's other "non-SAP" applications. This adds a further significant dimension to the MES philosophy. The MES is accordingly a collection of production-related services located outside the ERP application but which, via a suitable system platform, as described above, can be "woven" into a total application.

In this connection the META Group comments: "SAP NetWeaver is the first viable product to support the Enterprise Service Architecture. All of a company's applications should therefore be integrated as so-called services into SAP NetWeaver. The services will then be available to the user via a uniform architecture and interface. The user himself will then no longer actually see individual systems any more but only a solution which maps its processes."

According to a statement of the META Group, with its NetWeaver product SAP is in competition with IBM and Microsoft. These latter have the same strategy whereby efforts are being made, also on an operating system level, to launch a uniform platform onto the market.

The cost on the IT side for introducing applications should drop markedly in the future due to these strategies. For this to be possible, in NetWeaver SAP is providing a kind of development environment for its partners with which workflows and solutions are implemented: the Composite

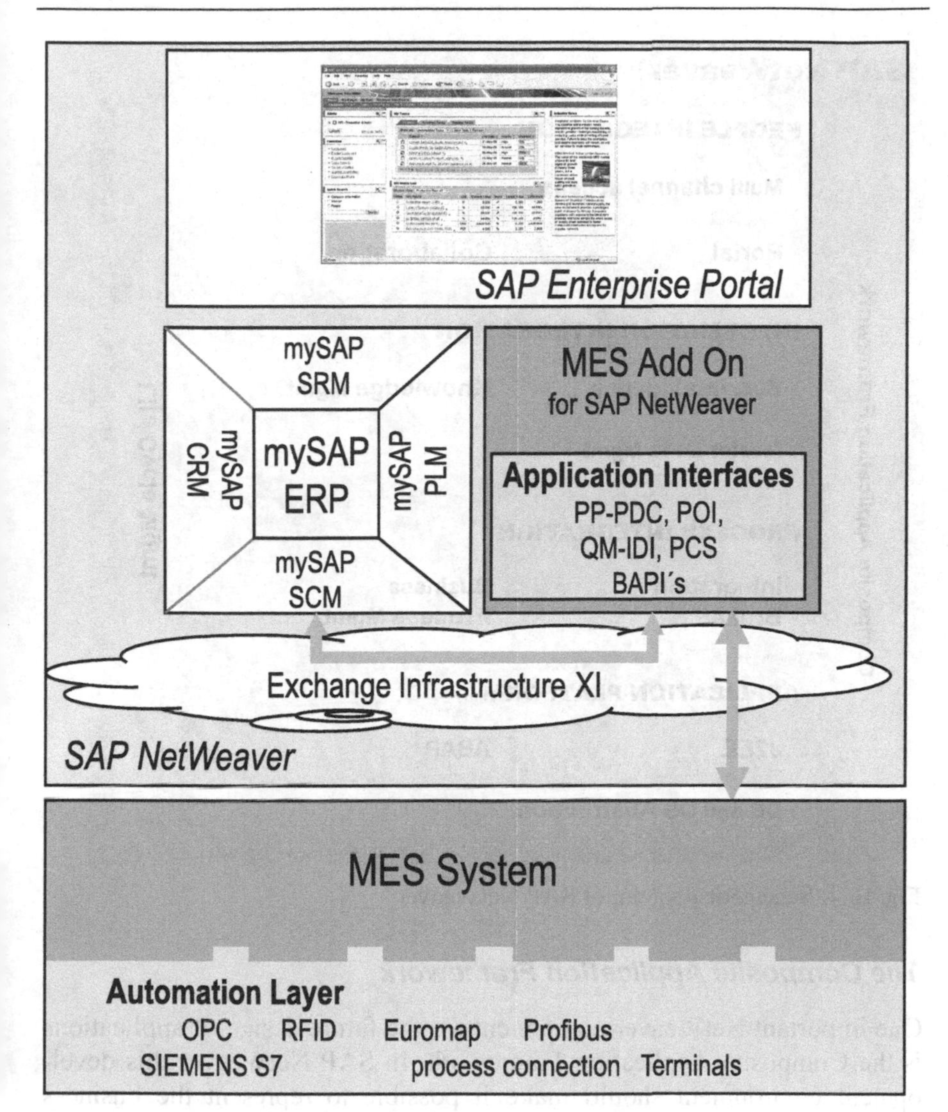

Fig. 10.3. Overview of integration scenarios in SAP NetWeaver

Application Framework. Among other things, it enables so-called cross-applications to be generated.

The exchange infrastructure (XI) is available for the technical connection of systems, including the MES system, to SAP NetWeaver. This is the communications level which all systems to be integrated in SAP NetWeaver use for communicating with each other.

For better basic understanding we will explain the terms "composite application framework" and "cross-application".

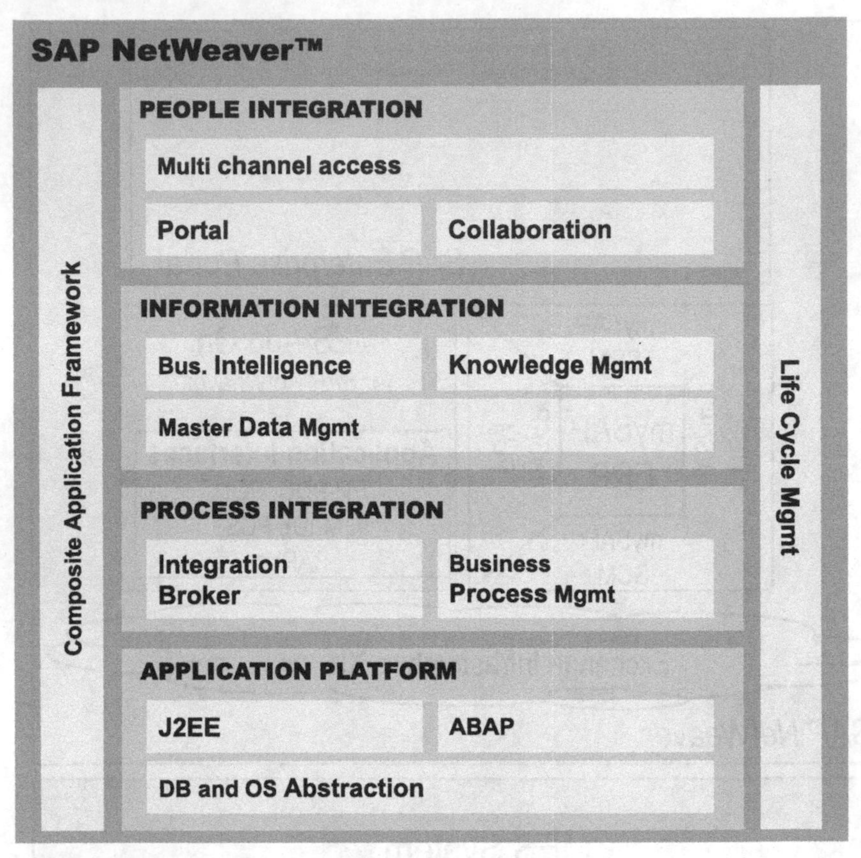

Fig. 10.4. Schematic diagram of SAP NetWeaver

The Composite Application Framework

One important NetWeaver component for the future design of applications is the Composite Application Framework. In SAP NetWeaver this development environment should make it possible to represent the business process as a workflow (guided procedures). This ideal state, provided it is attained, would put the manufacturing company in the position of being able to map all of its business processes in one system – the so-called SAP Business Suite – by means of SAP NetWeaver. Irrespective of which applications are in use, they will experience complete integration via SAP NetWeaver. In this connection META Group no longer talks about ERP consultants who introduce ERP systems but instead about "business process designers" who, with the aid of the development tools provided in SAP NetWeaver, design the processes of the manufacturing company directly in the system. For this reason SAP has integrated products such as Master

Data Management and Business Warehouse into NetWeaver. With Master Data Management, all of a manufacturing company's data should be managed centrally, irrespective of where the data are actually physically stored. This approach goes a long way beyond a "central" database since this data management application also contains the logical interdependencies and access paths.

Cross-Applications (xApps)

The SAP definition from the literature for the cross-application can be summarized as follows. A so-called cross-application links and aggregates data over and beyond the production landscape. Information from different processes, production lines and plant installations thus becomes more transparent. The application package contains preconfigured contents (the so-called iViews) as well as a direct integration with, for example, the Dashboards and Alert Management from mySAP ERP.

For this, the services for data access, analysis and business logic are used to connect up any sources of data in production desired such as, for example, solutions for production automation and control or maintenance applications.

The application package as put together allows all specialists involved in the value chain to work together on questions which crop up about the SAP Enterprise Portal and in any application of the mySAP Business Suite and to take steps to solve the problem.

What this means for a manufacturing company is that for it to be able to achieve optimum use of mySAP ERP in the future, an MES system must provide the requisite functions in this environment so that when used in conjunction with SAP, the user has a total solution for his production activities.

10.3.2 Interfaces with mySAP ERP applications

Basically with an implementation in the SAP environment the classic type of data exchange between two applications – for example, between SAP production planning (PP) or SAP merchandise management (MM) and the lower-level MES – must be possible.

SAP provides the individual SAP modules in mySAP ERP with interfaces which are also based on NetWeaver technology. The lower-level MES system transfers acquired data to the ERP via these interfaces and loads the data needed for the user's information and above all for controlling the MES from the SAP system.

The applications in mySAP ERP and in the MES system are linked via the application interfaces.

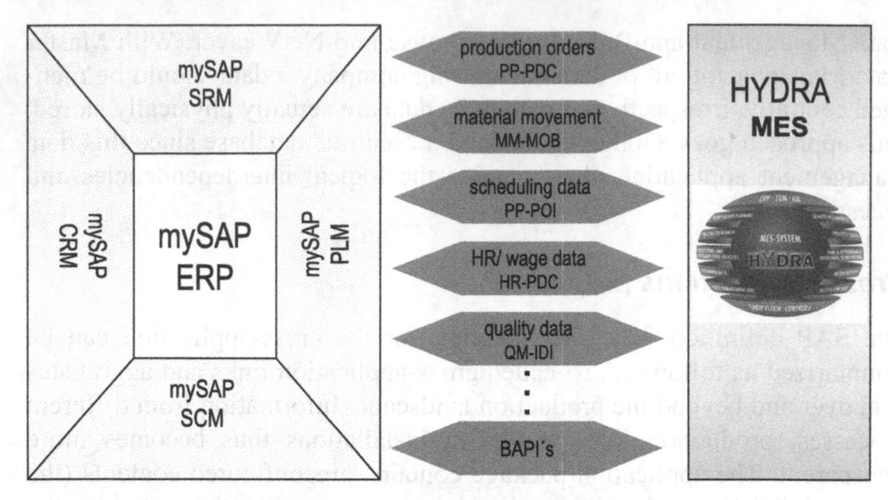

Fig. 10.5. Overview of application interfaces

With these interfaces, which are used at specific times to exchange information between the two systems, it is easily possible to connect together two processes which run completely separated in two different systems.

We shall now go on to deal with the most important interfaces with the ERP application. It is important that the MES supports the interface technology required. What is of decisive importance to the user, however, is the content of the data exchanged via the interface.

Plant Data Collection (PP-PDC)

This interface is traditionally used for exchanging status messages, master data and transaction data relating to production orders between mySAP ERP and the MES system. Via this interface the MES system reports time events or compressed data back to the SAP system in the form of time tickets. In this way mySAP ERP is supplied in real time with the production parameters. The usual operating mode for this interface in a lower-level MES is reporting via time tickets. Here totaled or condensed times and quantities are reported back on an order-related basis. Only in this way is preprocessing of the data possible and load is taken off the SAP system as far as the data quantity and the system performance of the subordinate system are concerned.

Production-Optimizing Interface (PP-POI)

With the PP-POI interface, mySAP ERP makes it possible for the MES system to read all master data and transaction data for material such as bills of material, work center information and also planning/production orders and warehouse stocks, and also to write modified information back to the production planning system. This function is of special importance when the MES is to perform shop floor control or reactive planning functions. In this case, the POI interface is used on the one hand when scheduled dates are sent to the control components of the MES in the form of milestones and on the other hand, following completion of plan optimization, when the MES reports the changed target dates back to mySAP ERP where the necessary "consequences" can be deduced from the date changes.

Inspection Data Interface (QM-IDI)

This interface is used in the field of quality management. The MES system reads via the interface the values required for quality assurance. This means that the entire inspection process can be carried out within the MES system and the measured values and quality information required for quality documentation reported back to mySAP ERP. It must, of course, be possible for this very comprehensive interface to be scalably operated by the lower-level MES. Depending on what form quality management takes in the manufacturing company, the full range of functions must be available, from simple measured data acquisition to full implementation and execution of the inspection and test plans specified by the SAP system.

Process Control System (PI-PCS)

A wide variety of control systems are in service in the process industry. Control scenarios range from fully-automated installations controlled by process systems to installations with a low level of automation which are mostly operated manually. So that these installations can be supplied with control information or can process messages received, SAP provides the so-called PI-PCS interface.

With the PI-PCS interface not only can control recipes be downloaded to the lower-level control system but process-related data can also be uploaded in the form of process messages. Order-related data are sent via the PI-PCS interface instead of via PP-PDC.

The following data are passed on in control recipes:

- Process and control parameters
- Plain-language instructions for the installation operator in partially automated or 100% manually operated installations
- Information about process messages to be sent back
- Process messages give information about

- The status of process orders
- The consumption and production of materials
- The status of resources
- Selected process events

Business Application Programming Interface (BAPI)

If processes in the company are to be "woven" more tightly together, data exchange between both systems will have to be more intensive. Should the processes need to be adapted customer-specifically, this will frequently mean adaptation being implemented in both systems (in MES and in SAP). As the production process becomes more flexible, the connection between the two systems thus becomes a burden. To enable more effective use of the systems, SAP offers its business application programming interface (BAPI) which makes it possible for sub-processes to be "detached" from SAP so that they can also be used by the lower-level MES system.

The lower-level MES will then behave like a virtual user in the SAP system. The advantage for the SAP user is that the MES is integrated by applying familiar and thus reproducible procedures within the SAP. This does require the MES to support and supplement the SAP functions in such a way that a wall-to-wall image of the production process is possible. For this reason, in implementation it is assumed that the MES vendor has not only the necessary process know-how relating to the manufacturing company but also the necessary familiarity with using SAP.

Customer-specific changes can be made in the MES by using this procedure without the need to make changes in parallel on the SAP side.

10.3.3 Integration of MES functions via the SAP portal

SAP's Enterprise Portal pursues the goal of making information from virtually any application available on the web. In addition to SAP there are other suppliers of portal solutions on the market, such as IBM, for example. However, by connecting the portal with NetWeaver and the ERP application itself, SAP enjoys competitive advantages with companies which are already using SAP applications.

Basically the SAP portal is thus an "application-neutral facility" which is intended to display not only SAP content but also MES content.

For all functional areas in the manufacturing company in which the user has no portal access or where there are key users or power users who need more powerful functions outside the portal, there will be an additional need for comprehensive MES applications which are available solely in the MES system itself.

The SAP portal offers the user of an MES system two interesting integration methods. The first is the “manufacturing intelligence dashboards”. These dashboards provide the user with navigation and alarm functions. The second is the so-called “business packages”: complete MES application packages which are normally a component of the MES system but which nevertheless run in the SAP portal.

Manufacturing intelligence dashboards

SAP or the SAP partner provides a manufacturing intelligence dashboard especially for evaluations and analyses in manufacturing companies. Based on the SAP NetWeaver technology platform it enables users to receive information from the MES system via preconfigured portals. The dashboards help production employees to make informed decisions more rapidly. SAP provides preconfigured roles for plant managers, production managers, maintenance managers and quality assurance representatives.

In today’s fast-changing world of work the availability of relevant and up-to-date information to group and team leaders and heads of department is of decisive importance.

A wide range of functionalities is available, including connection to production applications and data sources, the visualization and analysis of production data and other data, KPI and alert management, quality analyses, and event-based plant-SAP integration.

Via the KPIs (key performance indicators) production can access data about, for example, completion status, asset capacity usage and scrap rates. The so-called alert monitor sends warning messages (alerts) when there is a drop below safety stock levels, or when the scrap rate exceeds a specific limit value, or when an indicator for product quality points to a problem. The MES sends event notifications when, for example, there is a change in the order status, when a system fails or when there are quality problems. This information is then shown on the manufacturing dashboard.

The main application in the manufacturing environment is on the management level where current information is made available. Employees in production need detailed up-to-date information and thus need access to detailed MES functions.

For this the dashboard provides the important drill-down function. This allows production employees to “branch off” directly into their customary MES system environment. A genuine integration into the SAP user interface thus takes place without functions which are already present in the MES needing to be “reprogrammed” into mySAP ERP.

Business packages

SAP offers various applications and services in the form of user- and task-oriented business packages. In addition SAP partners can offer their own business packages in the SAP portal. They are based on SAP products and applications from MES partners. Essentially the packages in the manufacturing environment include information and workflows for the management level in production.

The business packages are a collection of so-called iViews (interactive views). Interactive views are independent software modules which insert themselves into the SAP Enterprise Portal and thus provide the user with all basic features (such as role orientation and the authorization concept). The iView thus offers a way of forwarding information (content) directly from the MES system to the portal user or recording it in a predefined workflow. The iView itself is however a constituent part of the MES system.

10.4 Support for SAP's Adaptive Manufacturing initiative

At SAP the term Adaptive Manufacturing has since 2005 primarily stood for activities which should make it possible for the mySAP ERP user to use MES functions.

With its associated partner initiative, SAP is thus informing its customers that for an optimized use in production mySAP ERP can be supplemented with an independent MES system or in the case of some industries must be thus supplemented. SAP thus creates for itself a clear market advantage since the SAP user has functions from executable MES systems at his disposal and the SAP application can thus adapt itself to the sometimes very heterogeneous manufacturing environments of different branches of industry.

With this initiative SAP is in addition pushing partnering and is linking this initiative to use of SAP NetWeaver. Qualified partners are being sought who can offer SAP customers a powerful MES system. In order to clearly indicate to SAP customers as well that these partners are qualified, SAP offers these partners the opportunity of acquiring the "Powered by NetWeaver" certificate.

The following criteria are relevant to the manufacturing company selecting an MES system to run under SAP.

10.4.1 Scalability of the MES solution

Before the necessary flexibility can be achieved for an application, the architecture of the MES system must ensure that an MES is scalable beneath mySAP. This also applies in the case of installation in a manufacturing company since the requirements made of an MES will be entirely different, depending on the production area. In production areas with a considerable level of automation, such as injection molding, the emphasis is on the technological interfacing of machines, starting with data acquisition of quantities or malfunctions and process parameters and including the transfer of machine settings. In a production area which makes intensive use of labor, such as the assembly line, what is important is registration of times and displaying screen information such as assembly instructions. These two applications require different data acquisition technologies.

In addition the production process itself must be in a process of permanent self-adaptation, First, as the result of technical optimizations of processes but Second also due to ever more dynamic changes to the product itself combined with continually declining lot sizes.

The dynamics of the production process in a company means that the MES system must be able to adapt itself and thus be scalable technically:

- Dynamic expansion capability of the IT infrastructure due to the systematic use of standardized components as operating system, database, and network technology
- Support for interfaces commonly used with the SAP applications (see Sect. 10.3.2)
- Integration into SAP integration scenarios in order to make possible a need-oriented distribution of information within the company (see Sect. 10.3.3)
- User-adaptable MES user interface for generating the user's own evaluations in production

10.4.2 MES for horizontal integration

To ensure that an MES system can guarantee future-proof service within a manufacturing company, functions must be available which secure horizontal integration for the application. Basically this means that components for:

- Order data acquisition,
- Machine data collection,
- Control station,
- Material and production logistics,
- Staff work time logging,
- Performance-based wages,
- Quality assurance,

– Access control,
– DNC,
– Process data,
– Tool and resource management

must be present before a manufacturing company can be offered a global solution covering the full scope of the application.

10.4.3 Interfacing on the machine and control levels

Another decisively important property with which the MES ensures that mySAP ERP can be integrated into different production organizations and branches of industry is its support for all common technologies on the automation level. In addition to using standard components, the MES system must be able here to interface with even very individually programmed process control systems, data collection systems or even with machines and machine groups which continue to be in service with many manufacturing companies. Only in this way can a wall-to-wall solution be found for the company and a new set of island solutions be avoided. The MES should on the contrary ensure that existing island solutions in production can be integrated into the MES system and a homogeneous interface with the SAP system be thus created.

An MES must provide the SAP user with the following common technologies and data acquisition standards:

RFID: Radio Frequency Identification

While in past years the use of bar code labels for identifying material, personnel and order documents has determined in-process data acquisition, the RFID method is still gaining in popularity. Since the technology has now become very inexpensive and due to the fact that the data carriers can be written within the manufacturing sequence, RFID promises a very wide range of applications such as, for example, tracking materials by using the data storage medium as "electronic" labels. Naturally this depends on the MES being able to provide the corresponding required application for this technique.

OPC: object linking and embedding (OLE) for process control

The OPC communication specification, which is now supported by a large number of manufacturers of production machinery and installations in the most varied branches of industry, is currently well on the road to establishing itself as a full-coverage standard. For this reason it is essential that an MES is able to use this way of reading machine, quality and process data from machines and machine groups. In addition, control information (so-called settings data) must be conveyed to the production equipment from the ERP via the MES in some cases on an order- or product-related basis.

As has already been mentioned, this functionality must be similarly supported above all in the field of plastic manufacturing standards such as Euromap E63 or manufacturer-specific communications protocols.

FDA and GAMP4 conformity

In addition to the technical standards already mentioned, the MES is also of decisive importance in the implementation of U.S. Food and Drug Administration (FDA) quality standards and the GAMP4 (good automation manufacturing process) regulations derived from them as also the corresponding European standards. This quality standard primarily requires of food and pharmaceutical product manufactures that they validate the ERP and MES systems used for production control and in production. In addition to the quality requirements which are made of manufacturers of the ERP and MES software, this software must also ensure that the safety requirements prescribed in the FDA's Standard 21 CFR Part 11 are satisfied. These are for the most part functional safety requirements which the software must meet. Today the use of an MES application for a manufacturing company which wishes to obtain an FDA-conformant validation is a basic requirement if the corresponding costs are to be kept within a reasonable framework.

10.4.4 Examples of the integration of MES and mySAP ERP

In the SAP environment in particular, it is important for the benefits of application to the manufacturing company that the technical requirements be separated from the actual benefits of using it. To make this clearer, the most important of the aforementioned possibilities will be summarized below in an application example. The idea behind this presentation is to compare the indicated technological possibilities with the corresponding application benefits.

This simple example concerns a wall-to-wall integration scenario involving, on the one hand, the MES system HYDRA from MPDV and, on the other hand, mySAP ERP.

The MES has a complete order data and machine data collection capability. To enable the MES to carry out order data acquisition and shop floor control it is interfaced with mySAP ERP via the application interface PP-PDC. As has been described in Sect. 10.3.2, order inputs are transferred as a stock of orders to the MES via the download function.

User-specific information is read directly in mySAP ERP. In the present example, the MES manufacturer employs a module he has developed himself and which SAP has certificated with the "Powered by NetWeaver" label. The order data are registered, in some cases aggregated, and then sent back to mySAP ERP via PP-PDC. Material movements are also registered in the MES system, with the work in process (WIP) materials not only being registered in the MES but material movements visualized. Messages are sent to SAP materials management (MM) only at the predefined milestones (lowest production levels). This means there is demarcation from SAP materials management (MM) and from so-called transportation management and thus on the system side an optimum supplementation of SAP by the lower-level MES. Should a machine standstill or operator intervention make a maintenance measure necessary, the corresponding maintenance job will be initiated in the PM module directly by using a BAPI. Whether registration of the PM order, in other words, aggregation of the times is effected by an MES function (because, for example, a machine data collection function is already running at the work centers or machines via the MES system) or whether these times are input manually via mySAP ERP functions depends solely on what IT infrastructure is present or on the user's organization. What is important for the flexibility is that the system configuration implemented and shown here supports both alternatives alike as options.

Alongside the classic functions of an MES system, in this example the SAP user has the following additional functions at his disposal and in this example they have been realized on the basis of the NetWeaver technology.

Manufacturing cockpit

Since machine data collection is now running almost completely as an application in the MES but certain information is to be available throughout the company, this information is offered to users at the SAP interface. Depending on what form the SAP system takes, this should take place at the SAP GUI or in the SAP portal (see Sect. 10.3.3). In the present example the user is provided in the SAP portal with a status overview of all machines, machine groups and work centers.

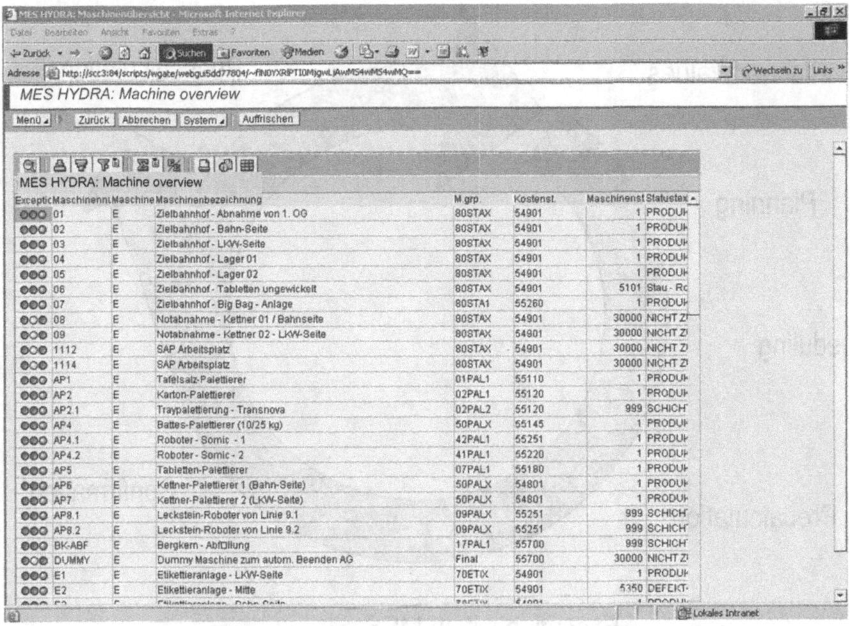

MES HYDRA: Machine overview

Menü Zurück Abbrechen System Auffrischen

MES HYDRA: Machine overview

Exceptic	Maschinennr	Maschine	Maschinenbezeichnung	M.grp.	Kostenst.	Maschinenst	Statustex
	01	E	Zielbahnhof - Abnahme von 1. OG	80STAX	54901	1	PRODUk
	02	E	Zielbahnhof - Bahn-Seite	80STAX	54901	1	PRODUk
	03	E	Zielbahnhof - LKW-Seite	80STAX	54901	1	PRODUk
	04	E	Zielbahnhof - Lager 01	80STAX	54901	1	PRODUk
	05	E	Zielbahnhof - Lager 02	80STAX	54901	1	PRODUk
	06	E	Zielbahnhof - Tabletten ungewickelt	80STAX	54901	5101	Stau - Rc
	07	E	Zielbahnhof - Big Bag - Anlage	80STA1	55260	1	PRODUk
	08	E	Notabnahme - Kettner 01 / Bahnseite	80STAX	54901	30000	NICHT Z
	09	E	Notabnahme - Kettner 02 - LKW-Seite	80STAX	54901	30000	NICHT Z
	1112	E	SAP Arbeitsplatz	80STAX	54901	30000	NICHT Z
	1114	E	SAP Arbeitsplatz	80STAX	54901	30000	NICHT Z
	AP1	E	Tafelsalz-Palettierer	01PAL1	55110	1	PRODUk
	AP2	E	Karton-Palettierer	02PAL1	55120	1	PRODUk
	AP2.1	E	Traypalettierung - Transnova	02PAL2	55120	999	SCHICHT
	AP4	E	Battes-Palettierer (10/25 kg)	50PALX	55145	1	PRODUk
	AP4.1	E	Roboter - Somic - 1	42PAL1	55251	1	PRODUk
	AP4.2	E	Roboter - Somic - 2	41PAL1	55220	1	PRODUk
	AP5	E	Tabletten-Palettierer	07PAL1	55180	1	PRODUk
	AP6	E	Kettner-Palettierer 1 (Bahn-Seite)	50PALX	54801	1	PRODUk
	AP7	E	Kettner-Palettierer 2 (LKW-Seite)	50PALX	54801	1	PRODUk
	AP8.1	E	Leckstein-Roboter von Linie 9.1	09PALX	55251	999	SCHICHT
	AP8.2	E	Leckstein-Roboter von Linie 9.2	09PALX	55251	999	SCHICHT
	BK-ABF	E	Bergkern - Abfüllung	17PAL1	55700	999	SCHICHT
	DUMMY	E	Dummy Maschine zum autom. Beenden AG	Final	55700	30000	NICHT Z
	E1	E	Etikettieranlage - LKW-Seite	70ETIX	54901	1	PRODUk
	E2	E	Etikettieranlage - Mitte	70ETIX	54901	5350	DEFEKT-

Fig. 10.6. Machine overview on the web

Manufacturing Master Data Control (MMDC)

MMDC stands for a continuous process of improving master data. It is important, especially for flexible production processes which are subject to constant change, for permanent comparison to be made between the planned required values and the actual values. The classic approach is for this target/actual comparison to be supported order-specifically either in the MES or in the ERP by the corresponding analyses. These analyses are however used exclusively for classic statistical cost accounting. In many cases what is missing entirely is feedback of this information in order to improve the planned values. These plan inputs form the basis not only of production planning but also of corporate planning as a whole. The job of MMDC is not only to support this control loop but also to make it a mandatory procedure for users.

With the aid of MMDC a control process is set up which for all orders collectively compares the plan inputs from the work plan and the improved target values obtained from the actual data. To secure the control loop, deviations of this kind are displayed as a sort of to-do list in an overview for the user.

The appropriate employee must work through this list assigning every deviation to its cause.

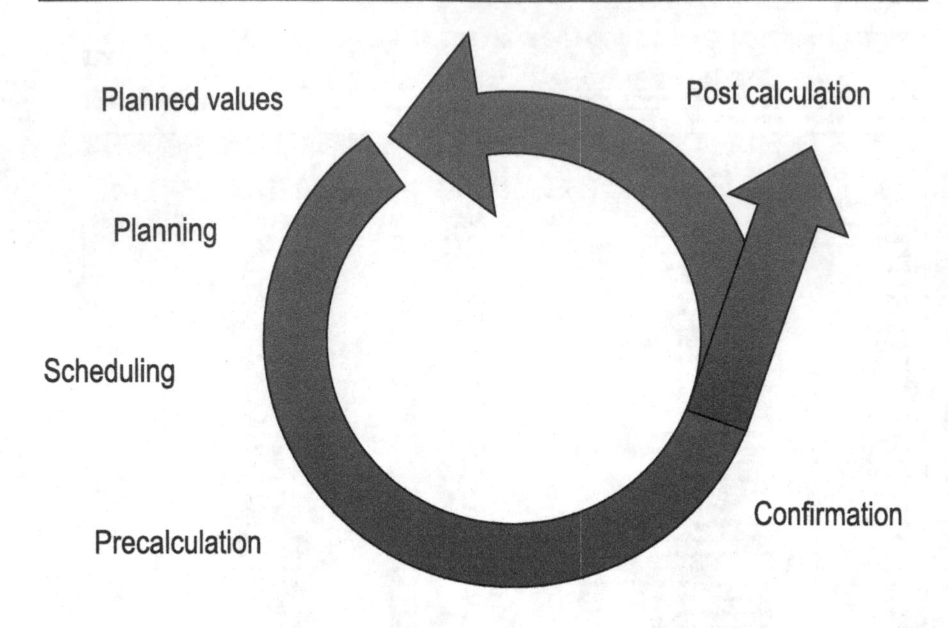

Fig. 10.7. Control loop for optimizing the manufacturing master data

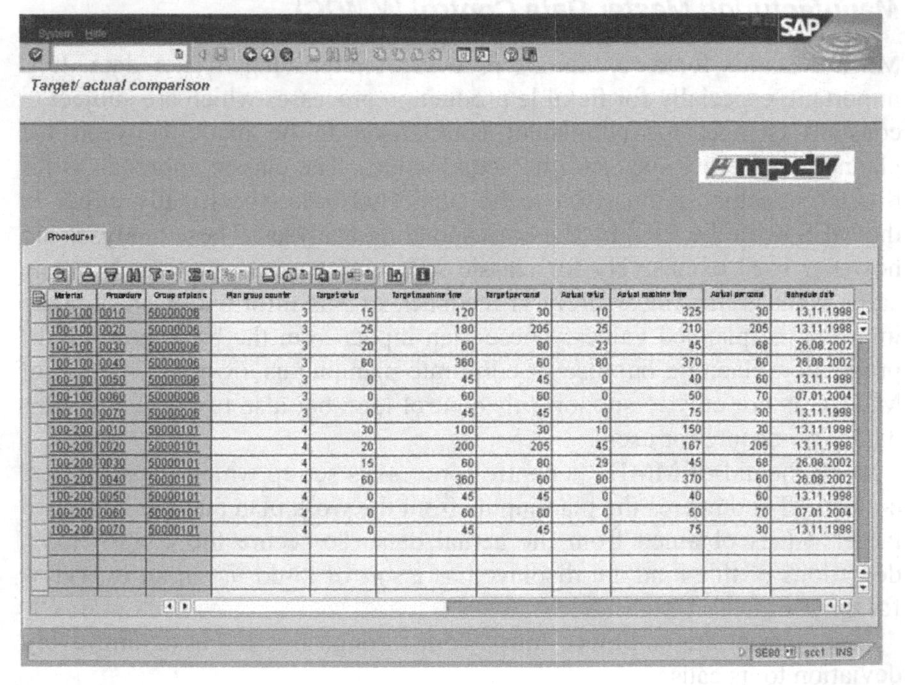

Material	Procedure	Group of plans	Plan group counter	Target setup	Target machine time	Target personal	Actual setup	Actual machine time	Actual personal	Schedule date
100-100	0010	50000006	3	15	120	30	10	325	30	13.11.1998
100-100	0020	50000006	3	25	180	205	25	210	205	13.11.1998
100-100	0030	50000006	3	20	60	80	23	45	68	26.08.2002
100-100	0040	50000006	3	60	360	60	80	370	60	26.08.2002
100-100	0050	50000006	3	0	45	45	0	40	60	13.11.1998
100-100	0060	50000006	3	0	60	60	0	50	70	07.01.2004
100-100	0070	50000005	3	0	45	45	0	75	30	13.11.1999
100-200	0010	50000101	4	30	100	30	10	150	30	13.11.1998
100-200	0020	50000101	4	20	200	205	45	167	205	13.11.1998
100-200	0030	50000101	4	15	60	80	29	45	68	26.08.2002
100-200	0040	50000101	4	60	360	60	80	370	60	26.08.2002
100-200	0050	50000101	4	0	45	45	0	40	60	13.11.1998
100-200	0060	50000101	4	0	60	60	0	50	70	07.01.2004
100-200	0070	50000101	4	0	45	45	0	75	30	13.11.1998

Fig. 10.8. Target/actual comparison of the plan data

Tracking targets in manufacturing: the manufacturing scorecard

The manufacturing scorecard is a method developed by MPDV Mikrolab GmbH for system-supported definition of targets in production and tracking the degree to which they are being or have been attained. In the SAP environment this is a typical example of the application of key performance indicators (KPIs) in a manufacturing intelligence dashboard. The target in question from the manufacturing scorecard module is saved to the dashboard. An overview of the defined targets and their corresponding fulfillment levels is displayed in the so-called KPI watch list on the dashboard. If detailed information needs to be accessed, branching off into the MES system is recommended.

KPI Watch List

Choose View: Morning Check | Define Views | Personalize KPIs

Status	Key Figure	Unit	Current Value	Trend	Target Value	Difference	Difference [%]	Time Stamp
	Capacity Utilization - Chicago/All	%	25.200	→	91.000	65.800-	72.31-	29.04.2004 16:38
	Capacity Utilization - Oswego/All	%	0.000	↘	90.000	90.000-	100.00-	28.04.2004 14:32
	Confirmed Quantities per Order	%	1.000	↘	100.000	99.000-	80.00-	28.04.2004 14:41
	Coffee Machine Level	L	1.400	↗	1.500	0.100-	6.67-	28.04.2004 14:40
	Target/Actual Analysis - Chicago/All	%	2.700	↘	1.000	1.700	170.00	17.04.2004 02:00

Refresh

Achievement of objectives:
e.g. process degree,
Efficiency factor

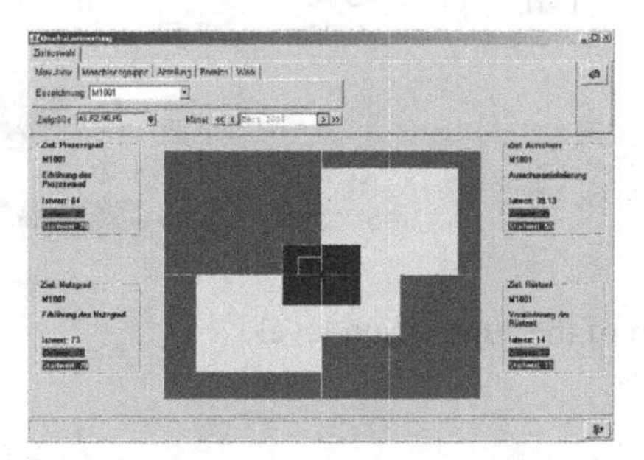

Fig. 10.9. Monitoring target achievement by means of the manufacturing scorecard

Interfacing with the automation layer

As regards interfacing with machines and process plants, the MES in this example supports modern technologies such as OPC (see Sect. 10.4.3). What is of decisive importance is that the MES implements standardization at the machine interface for the process level in order to be able to integrate what is generally a heterogeneous control environment and without having to implement a large number of proprietary interfaces with the SAP system.

In the present example, along with the product the manufacturer has supplied the corresponding process communication controller which makes technologies available for the virtually standardized interfaces (as described above) with the SAP system and with the machines and process plants. These technologies include:

- OPC
- Direct connections to control systems such as Siemens S7
- Digital I/O contacts for quantity counting and fault signal acquisition
- Drivers for field bus systems, such as Profibus
- Drivers for machine control networks
- Direct connection to weighing devices or complete weighing systems

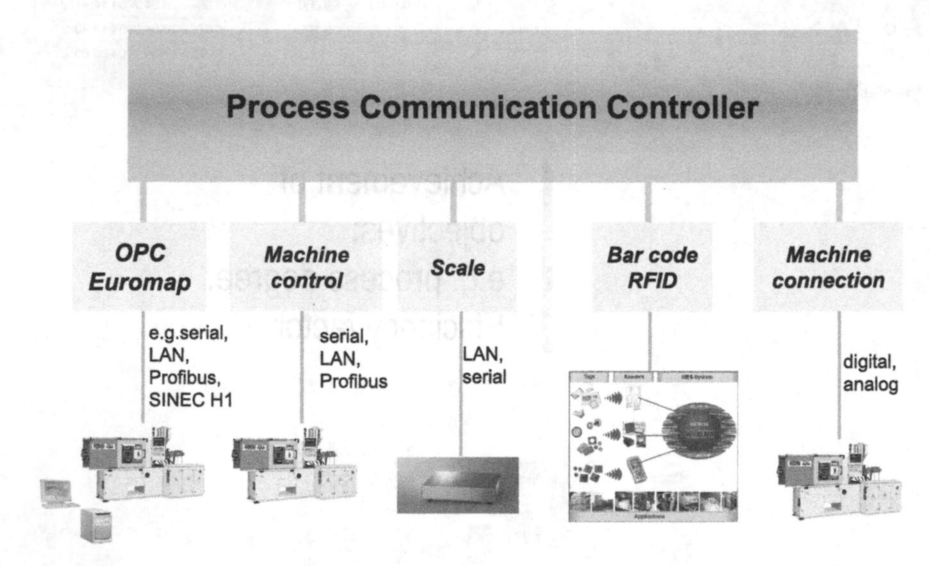

Fig. 10.10. Connection of the automation level

10.5 Summary

The MES concept is becoming increasingly important in the SAP environment since SAP is clearly communicating to users of its products that mySAP ERP is best integrated into the increasingly flexible world of production via an independent MES system.

With SAP NetWeaver, SAP intends to provide the technology platform required for this. Some of the technologies are still under development and market penetration will take some time yet. Until then it is however a matter of using existing technologies to meet the present demands of users and of using an MES which, although it will support the future-proof NetWeaver application, can even now demonstrate its powerful capabilities in the application while making use of the current integration technologies available.

The main thing for the SAP user is that, in addition to the technological incorporation of the MES system into the SAP world, the MES has the necessary flexibility with respect to the application and also supplies the complementary functions for production required by the mySAP ERP. Also of decisive importance here is that when installing the MES or mySAP ERP the consulting partners have the requisite process know-how relating to production procedures in the company.

11 MES in plastics processing

11.1 Special features of the plastics industry

The plastics industry, especially its discrete forms of manufacturing, has for historical reasons been a pioneer in the fields of production data acquisition, detailed planning and quality assurance. The networked philosophy of an MES system gives considerable advantages. In this important branch of industry machines and tools are at a high technological level and the factories are fitted out with modern equipment.

Processes used in this industry include injection molding, blow-molding, extrusion and also special types of processes derived from these. The plastics production department is usually connected to other automated production areas. In many cases the final products consist of a combination of the components plastic, metal and rubber. Here the advantages of an MES system can also be leveraged in other branches. Punching and compression molding and also all cycle-controlled automatic machines can be integrated into the system without any problems and this in turn delivers further advantages for the entire production segment.

As to the question of increasing economic efficiency, in many cases the main focus of discussion is investment in the machine park and in tooling. A closer examination will however reveal that it is considerably more economically efficient to improve the planning and to precisely record and evaluate all reasons for malfunctions and scrap than to attempt to achieve a higher return on investment by technological means. The objective is to achieve better utilization of the existing machine park and improve quality, and by taking this road reach higher efficiency. An MES system with its functions of the electronic control station, production data acquisition, machine data collection, tool and material management brings transparency to production and increases economic efficiency. In this regard Sect. 11.11 provides a detailed calculation which takes into consideration the various influencing factors.

Use of MES systems will always be particularly effective when the following conditions are met:

- Possibility of automatic data acquisition directly at the machines
- Automatic operation

- Presence of a relatively large number of machines of the same type
- Manufacturing using high-investment machines and tools
- Operation 24/7
- Requirement for high utilization ratio and percentage utilization
- Production of high quantities
- Requirement for "just in time" delivery capability

All of the conditions just listed are met in the production of plastic components by injection molding. Important areas of application in the plastics sector include:

- Automotive industry
- Engineering components
- Packaging
- Medical technology
- Electrical industry

Products in the plastics sector are complex, produced in large quantities and must satisfy a considerable number of requirements relating to quality, value for money and delivery reliability. Since technologies, machines and tools have already reached a high standard, at this point only evolutionary improvement possibilities are available. It is therefore much more efficient to ensure with the aid of an MES system that production is running with a high utilization ratio, a low scrap rate and high quality.

11.2 Usable MES modules

All MES modules can be used in the plastics processing industry. The modules will be referred to below using the terminology provided in VDI (Association of German Engineers) guideline VDI 5600.

Table 11.1

VDI guideline	Plastics
Detailed planning and control	Control station
Material management	Material management
Operating resources management	Tool management
Personnel management	Staff work time logging
Data acquisition and processing	Production data acquisition
Interface management	Interfaces with ERP/PPS/payroll
Performance analysis	Analyses
Quality management	Quality assurance
Information management	Management information

An integrated deployment of an MES system delivers the benefits in particular of the interactions between the different modules which permit data, once acquired, to be processed in different modules and the corresponding action taken on this basis. This will be demonstrated here via the example of one of the responses to scrap processing.

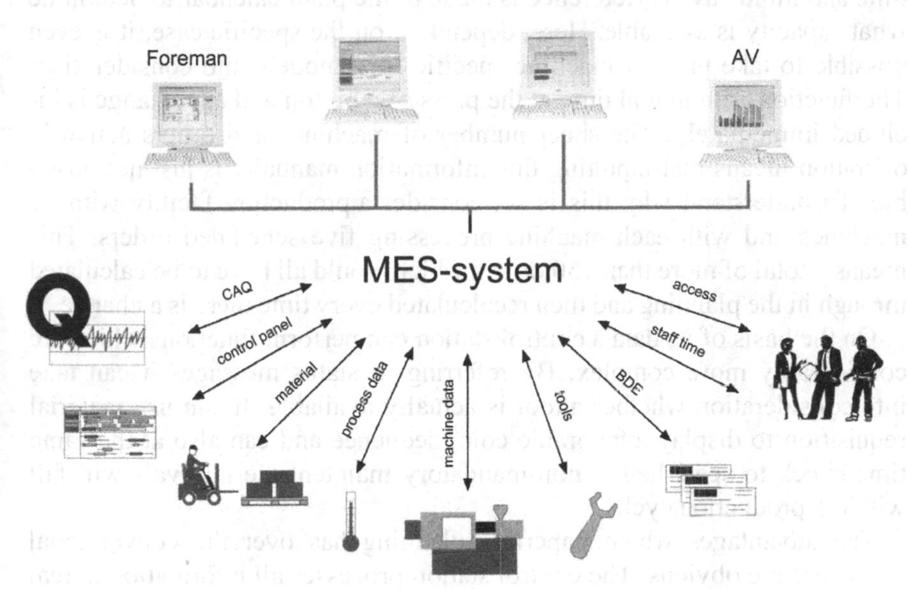

Fig. 11.1. MES modules

The scrap is detected and registered in the system quantitatively together with the relevant scrap reason. In the QA system this scrap reason is input into the analyses and statistics and on this basis the necessary action is taken. The quantity is reported to the control station and from here the corresponding secondary finishing is arranged and all consequences for subsequent orders arising from this extended processing time are displayed automatically. Material management checks to see whether the corresponding amount of material is available for secondary finishing. In the tool management domain the faulty parts are classified under the corresponding tool: this makes it is possible to investigate to what extent the tool was responsible for the scrap. Of particular importance is an economic analysis of the faulty parts which is sent via the interface with the ERP/PPS system. This puts the ERP/PPS system in a position to determine under real-time conditions the financial expense caused by the quality problem.

11.3 Control station

Use of a control station is particularly effective in plastics processing. From the requirements in the master data it is possible to calculate exactly the duration of the job on the basis of the data for quantity, set-up time, cycle time and mold cavity. Reference is made to the plant calendar to determine what capacity is available. Here, depending on the specific case, it is even possible to take part- or machine-specific shift models into consideration. The functions run in real time at the press of a button and any change is included immediately. The sheer number of machines and orders active in operation means that inputting this information manually is just not possible. To understand why this is so, consider a production facility with 30 machines and with each machine processing five scheduled orders. This means a total of more than 150 orders which would all have to be calculated through in the planning and then recalculated every time there is a change.

On the basis of its data a control station can perform functions which are considerably more complex. By referring to status messages it can take into consideration whether a tool is actually available. It can use material requisition to display a favorable color sequence and can also at the same time check to see whether non-mandatory maintenance intervals will fall within a production cycle.

The advantages which paperless planning has over the conventional procedure are obvious. The control station processes all information in real time and for this reason the planning background will always be up to date.

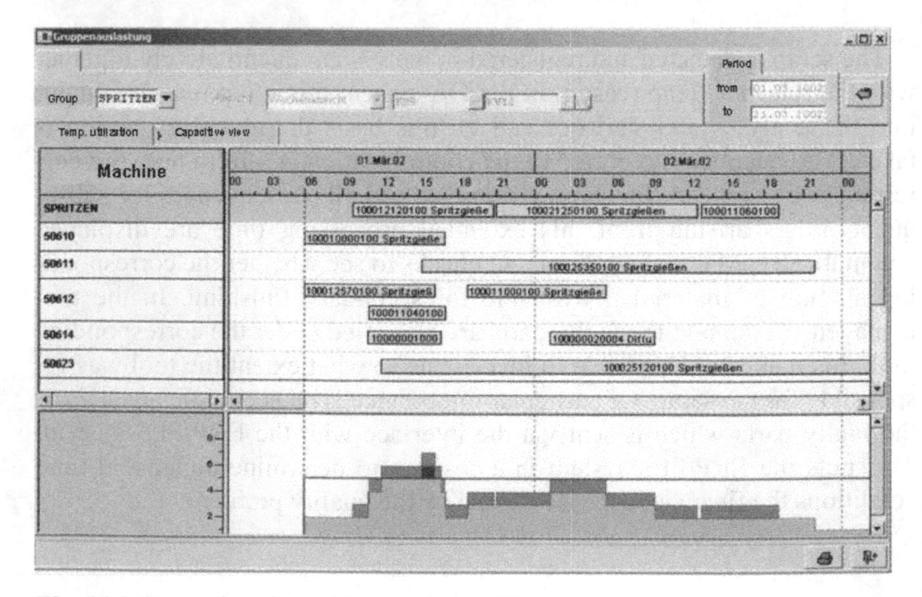

Fig. 11.2. Control station with capacity profile

Graphic support makes planning at the control station considerably more simple than working with a great quantity of paper documents which all have to be laboriously gathered together every time replanning is carried out and then rewritten entirely. Since the control station has the ability to make suggestions and, if the situation so requires, can also issue warnings, incorporation into planning is made much easier. This advantage is particularly useful when there are personnel changes or if a planner becomes unavailable at short notice.

The control station displays the sequence of orders in the form of a bar diagram. The corresponding loading graph shows at what time free production capacities are available and where overloading makes replanning of operations necessary.

11.4 Acquisition of machine and production data

Machine data collection (MDC) is of special importance to planning. Since any planning is only as good as the actual circumstances on which it is based, one of the essential foundations of neat planning is the automatic provision of quantities. Since precise figures for the number of parts produced are supplied fully automatically by the PDA part of the MES system, planning has a high level of actuality and of quality.

Planning benefits extensively from MDC. In addition, more information is collected manually via PDA (production data acquisition) and input at terminals in the factory. This includes set-up and servicing times and also reject numbers. This direct acquisition at the terminals is more certain, more up-to-date and associated with considerably lower costs than manual records which then have to be forwarded and subsequently input into EDP systems. Inputting immediately creates a real-time reference and, if so desired, even a reference to an individual can be created via confirmation by card reader. The electronic method of acquiring and transferring data is far superior to manual handwritten records as regards accuracy and efficiency. Companies which still work manually are usually aware of how weak the data framework is which forms the basis for discussions about utilization ratios and scrap rates.

Data acquisition stations may be either special terminals at the machines or PC computers with a Windows interface. These terminals can be used as individual or group terminals. They have a graphical user interface and functions are launched by touching on-screen symbols (touch screen). Corresponding to the application in question, operation can be configured to suit different requirements.

Reader stations for the identification of operators, materials or batches can be connected. Here the full range of options can be used (magnetic card, bar codes, chip card, RFID, and so on).

11.5 Connecting the injection-molding machines

There is no doubt that machines used in the plastics processing industry can be regarded as the ideal partners for machine data collection. There are various types of integration and here it is an advantage that basically any machine, irrespective of model, manufacturer or age, can be connected in a straightforward manner. The following possibilities are available:

- Cycle signal
- Cycle plus additional further digital signals
- Host computer protocols via field bus systems
- Euromap 63 as defined standard for injection-molding machines
- OPC as machine integration standard in the Windows environment

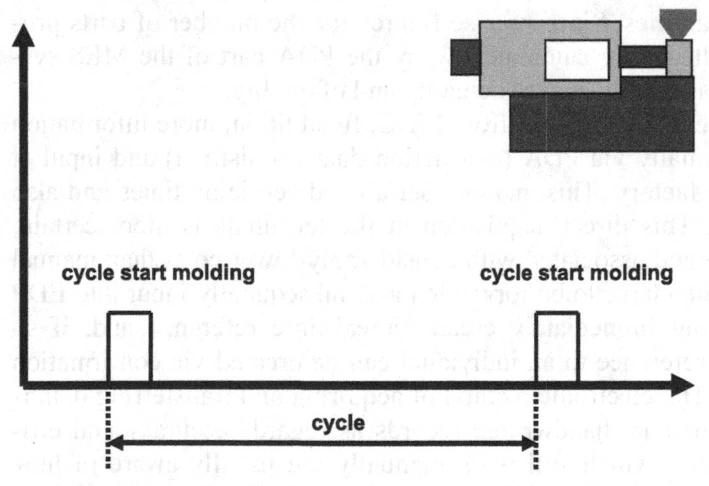

Fig. 11.3. Connecting injection-molding machines via cycle signal

One very simple method with a very favorable cost/benefit ratio is connection via the cycle signal which is available at every machine. This signal permits automatic detection of the following:

– Machine running/stopped	Signal received/not received
– Machine running too fast/too slow	Comparison with required cycle time
– Quantity produced	Cycle x mold cavity

In parallel with this, the reasons for standstills and numbers of scrap are acquired and recorded at the terminals. The status of the machines is thus easily documented and the information is available in real time for the corresponding analyses.

This type of data acquisition is in all respects superior to the use of conventional aids such as pen and paper. Manually prepared records come nowhere near the quality and currentness of data automatically and unerringly acquired by an MES system. Not only that but the cost of manual data acquisition is considerably higher. In the present-day production environment no one would dispute the need to collect data from operations. In this regard let us quote a sentence from a conference paper:

> "Manufacturing without production data is like driving with a windscreen you can't see through."

11.6 Visualization and evaluation

The huge amount of information flowing into an MES system requires preprocessing to allow the operator to work comfortably with the system. For many application cases, visualizations and graphical evaluations are available. Alarm functions and status changes can be very prominently displayed by means of color coding. The color "green", for example, indicates that everything is proceeding within the permitted limits. "Yellow" indicates a warning function while "red" means that the permitted limits have already been crossed. The operator can plan his actions correspondingly and access further analyses for the critical cases. Pie diagrams, bar diagrams and even plotted graphs are available. The use of numerical outputs, which are controlled via database selection, allows faults to be traced back successively to their origin. A report generator is available for preparing one's own analyses and reports. By default the export function in Excel is also available, allowing specific reports to be prepared on this basis. Figure 11.4 shows a typical display.

The analyses are used for investigating the causes of faults and for documentation purposes. In addition it is however necessary to have production under control at all times. The graphical machine park is used for this and gives a view over production via the monitor screen. The shop and its machines are shown in graphic form with colored displays which not only indicate warnings but also show the corresponding statuses. The user can generate the display easily and also modify it when machines are retooled. Production can be viewed from any location via the network and via the internet. By clicking on a machine, more information can be obtained.

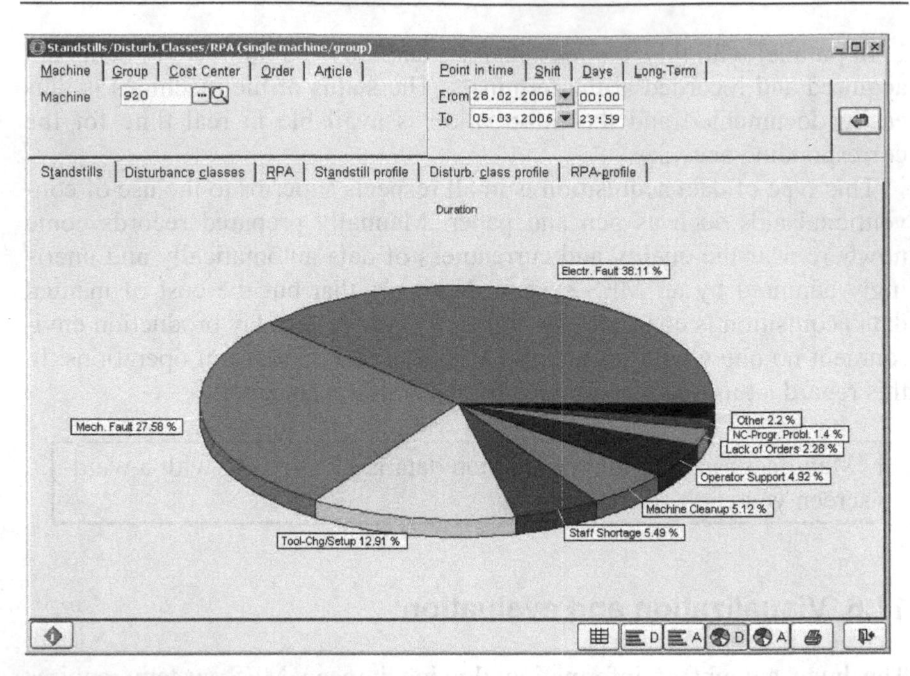

Fig. 11.4. Distribution of reasons for faults

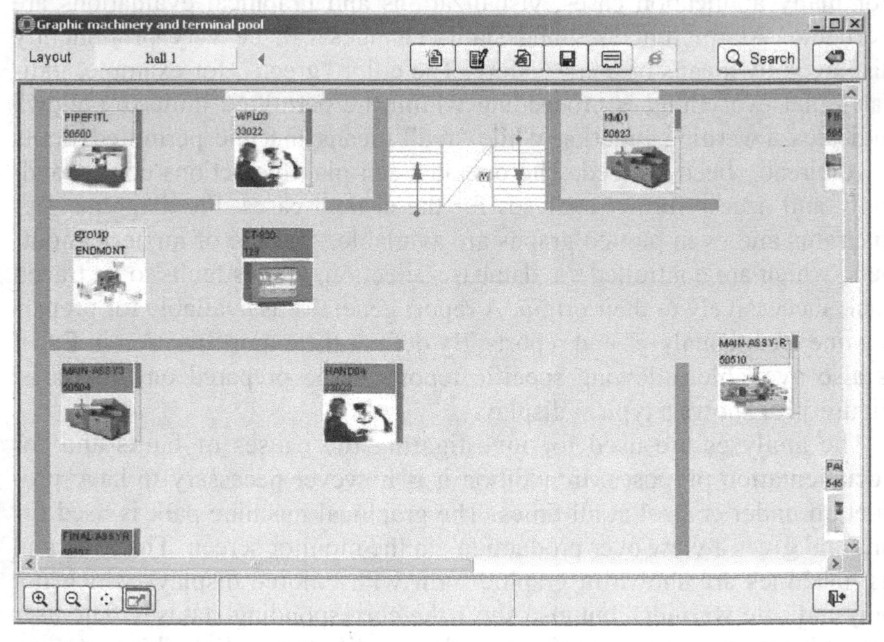

Fig. 11.5. Graphical machine park

The appropriate protective mechanisms are in place to restrict access to authorized users only. With this the possibility of making value processes measurably transparent is provided for the first time by an MES: only what can be measured can be changed.

During the course of the increasing orientation of work towards the process result there is a greater need for evaluation of the effectiveness of action taken. The result of this in practice is for the measured variables incorporated in operational target agreements to be presented in reviews not only as the levels which have currently been achieved but also in the form of historical curves. While statistics do not normally have an addressee, reviewing relates specifically to the employees and their performance. It is therefore generally true to say that a company cannot become process-capable until statistics have been globally replaced by reviewing.

The graphic machine park provides a rapid overview of the machines and of productivity. Simple program tools are provided to allow every operator to put together his particular machine shop.

11.7 Connection between quality assurance and process data

One particularly interesting application in the MES field ensures that quality assurance is connected to the process data which are obtained directly from the machine controller. It should be recalled at this point that in production important material properties come into existence during the process. In addition to basic properties of the molding compound, factors such as temperature, pressure, air humidity, drying level and so on play an important role. The quality assurance which follows can be set up on two foundations: First, by the objective characteristics which are registered or measured in a step which follows production and, Second, by the influencing variables which are documented by recording of the process parameters.

Monitoring the mold temperature in the case of thick-walled moldings may serve as an example here. Once the temperature falls below a critical value there is a risk of voids forming. Process data acquisition allows temperature values to be monitored and action taken immediately there is a deviation from set-point. A problem of this kind can only be analyzed at high cost using conventional measurement techniques or involving destruction of the piece. The situation is similar with safety-related moldings as regards the clamping force and its duration.

Process parameters can be recorded in all machines with a modern controller. This kind of monitoring of selected values can be a major factor in obtaining information which it would be very difficult or even impossible to obtain during the course of an assessment at a later stage. Regarding costs, automatic parameter monitoring methods are particularly advantageous. Since in many cases the interfaces for PDA and DNC are already present, this opens up a highly interesting possibility along the lines of:

"Make quality instead of checking quality."

The information in the MES system guarantees a production environment on which the quality needed today is based.

11.8 Tool making

There are two starting points for installing an MES system in the tool making department.

11.8.1 Using an MES system to monitor the maintenance intervals

Assignment of the production orders to a particular mold and machine combined with the gapless acquisition of all information ensures that the complete production history of a mold is recorded in the MES system right from the start. The required maintenance intervals are input as well and monitoring then runs automatically, providing advance warnings and graphic displays to show when a maintenance activity is necessary. In the system

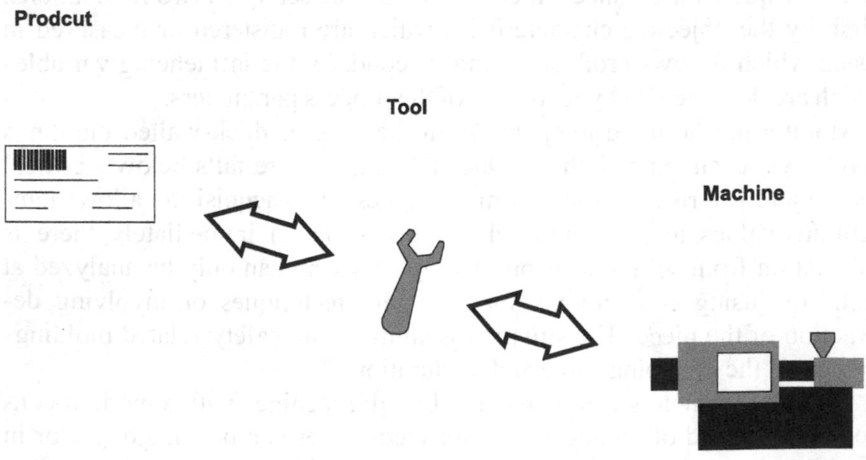

Fig. 11.6. Monitoring maintenance and service intervals

the operators must confirm that the maintenance has been carried out, thereby meeting the mandatory recording requirement and fully documenting the procedure. Repairs, overhauls and cleaning are also documented in the system and the corresponding information is available online in the network for everyone involved in the production process.

In the plastics processing industry, investment in tools and machines runs into tens of millions of euros. The personnel costs which have to be paid for documentation, maintenance and repair work are also high. Compensation costs or returns of items which, for example, arise in the automotive industry as soon as promised servicing events cannot be complied with or cannot be verified put profitability at risk. With manual management of this, the outlay is also immense and still does not ensure compliance with all requirements: a system of this kind holds too many sources of potential error.

Compliance with requirements cannot be depended on unless all information has been recorded in the MES system and events are automatically triggered via warning and execution actions.

11.8.2 PDA and control station in tool making

A high proportion of plastics companies have their own tool making department in which the molds are built and maintained. The machines in these departments also need to be scheduled and used efficiently. Due to tight deadlines it is particularly important here to make the best possible use of bottleneck machines. Whatever the case, monitoring of the times specified for mold construction is necessary if the cost situation here is to be kept under control. Not only machine times but also personnel times must be recorded and assigned to cost units. Since an MES system includes the modules of machine data collection and staff work time logging, it can be used in the tool making department.

On the basis of how the system has been installed in production it can be expanded to include tool making. Here the existing network is used and the existing master data can be used for parts and molds. Data acquisition at the machine tools can be automatic, or manual data acquisition terminals of the PC type are used, where the operator logs on or off.

11.9 DNC, batch tracking and mandatory in-process documentation

As part of documentation obligations a documentary record of all events throughout the full life cycle of a product is required. Due to the huge amount of data required and the requirements for immediate serviceability, this would be impossible without the use of automatic facilities. This is where we feel the benefit of the fact that all information relating to machine, tool and material is already available due to gapless PDA acquisition. By inputting the batch numbers a link is created on the material side to the batch and the duty of documentation is fulfilled for this part.

Communication between the MES system and the machines via interfaces can be used not only for transferring PDA and process data but also for sending settings to the controller. This brings savings in the handling process and, in addition, the individual production data associated with an order can also be fully documented.

The huge amount of time which would otherwise have been needed for manual documentation is saved completely while at the same time the quality of the data registered is considerably improved. It is easy to provide a complaining customer with a proof that all data subject to mandatory documentation really have been recorded. Customer inquiries in this regard can be easily satisfied at minimal cost by means of database evaluations. In

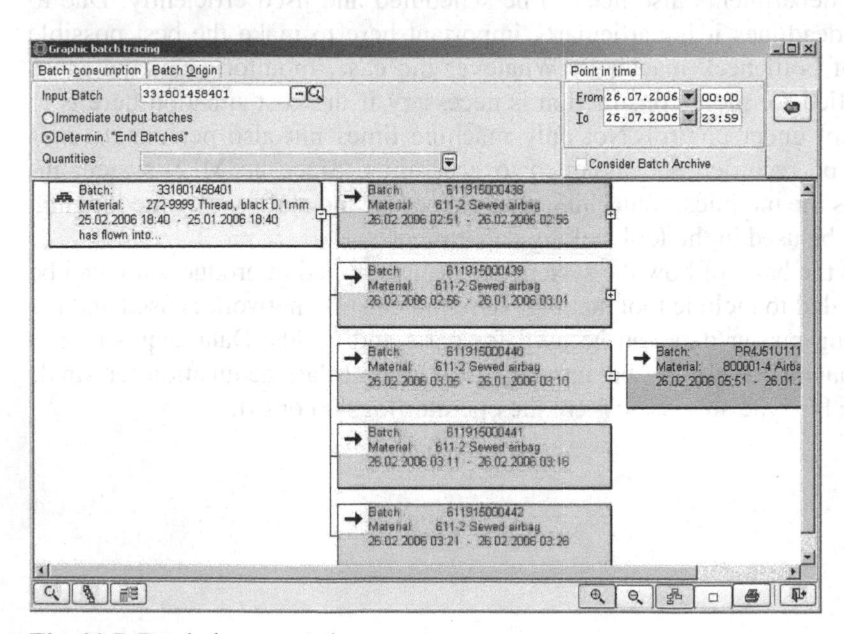

Fig. 11.7. Batch documentation

the case of quality problems, nonconforming parts can be easily identified by batch tracing and then quarantined.

Figure 11.7 shows the batches and lot numbers of which a product is composed. This allows tracing back to the material used and its corresponding lot number. Furthermore it is also possible to ascertain into which other lots the same material was input.

11.10 Management Information System (MIS)

Without an MES system an informative information structure is inconceivable. Only when all data about production times, downtimes and reject numbers are available in real time without omissions will management have an informative information background at its disposal and be able to make decisions on this basis. This calls for a powerful interface with the ERP/PPS system. Here the data are exchanged in both directions. There is a clear distribution of tasks between the two systems. The ERP/PPS system runs the bill-of-material explosion and generates the production orders for the individual work centers or work center groups. Here part numbers, quantities and the end of the order are specified as part of rough planning and transferred to the control station. Normally data transfer takes place once or twice a day. Detailed planning using graphic resources is carried out directly at the control station. This depends on the support of the PDA data. For example, the status of a mold is known in the MES system (free, scheduled elsewhere, under repair) and detailed planning can access this information in real time. Data transfer from the MES system to the ERP/PPS is responsible for the economic evaluation of the processes. When objective data are linked in real time to the financial evaluation from the ERP/PPS system, then you have a real management information system (MIS). The transfer cycle can be specified by the ERP/PPS system since all information is present in real time within the MES system.

It needs to be stated specifically at this point that planning and control cannot be performed by the ERP/PPS system alone. Two arguments need to be considered in this regard. First, the ERP/PPS systems are mostly not able to act in real time. With the machine-hour rates which apply today, even minutes have relevance to costs and decisions have to be taken very quickly indeed. On the other hand, the planning scenario of an ERP/PPS system always rests on a theoretical basis since a direct link to the production data is missing. This means that active control of operations is only possible to a very limited extent. This is why the use of an MES system is indispensable for a modern and effective factory.

11.11 Return on investment

Using an MES system in the plastics industry brings the following advantages:

- Better flow of information
- Production data with a high level of quality and accuracy
- Creation of the necessary transparency in all areas
- Uncovering areas of weakness
- Departure from paper-oriented production
- Shortening of lead times
- Reduction in scrap quantities
- Performance of mandatory in-process documentation
- Increasing productivity

What follows shows the improvement in economic efficiency secured when MES is used in an injection-molding company with 25 machines. Calculations are based on savings in four areas. The automatic acquisition at the machines of data relating to quantities and downtimes results in a drastic reduction in costs compared with the manual collection of data which would otherwise be necessary. This can be equated with savings of €25,000. The MES system increases the utilization ratio by means of online data and full transparency in production. In the experience of companies deploying a system of this kind improvements of 1–4% are known. For our purposes we shall assume an increase of 1% in the utilization ratio. In relation to the net product of 25 machines, this results in a saving of €30,000. Planning is considerably simplified and in the present case this results in personnel savings of €16,000.

Quantitative and qualitative data acquisition, registration and evaluation for all scrap results in a considerable reduction. If a reduction in scrap of just 0.4% is assumed, with respect to turnover this yields a saving of €15,000. Considerable savings arise from the fact that maintenance and servicing management can be handed over in their entirety to the MES system. Requirements are contained in the master data part and the current status is continuously reported by the MES system. All required actions are thus monitored automatically and displayed in graphic form on the screen. Assuming that 15 minutes per month must be devoted to each production resource, this yields potential annual savings of at least €9000.

With the values listed in this example, this means an annual saving of €95,000. The author will be pleased to supply on request a detailed breakdown with parametrizable influencing factors.

Economic efficiency of an MES system © WN-Consult		Unit	Value
	Working weeks per year	wk	53
	Working days per week	day	7
	Working hours per day	h	24
	Number of machines	pcs	25
	Number of molds active	pcs	60
	Number of peripheral devices	pcs	13
1	**Outlay on data acquisition and evaluation**		
1.1	Time for manual **data entries** per machine per day	min	7
	Quantities, scrap, downtimes, reasons for faults		
1.2	Time traveled, registered by shift all machines	min	91
1.3	Time for inputs, reports, analyses for all machines	min	45
	Time required per day	h	2.27
	Hourly pay rate for employees	€/h	30.00
1.4	**Total cost of data acquisition and evaluation**	**€**	**25,228**
2	**Loss of effectivity due to problems being detected too late or not at all**		
2.1	Standstills detected too late	%	0.5
2.2	Due to points of weakness going undetected	%	0.5
	Value added per year in molding shop	€	3,000,000
2.3	**Total loss of effectivity**	**€**	**30,000**
3	**Cost of manual planning**		
3.1	Set-up activities per machine per week	no.	3
	Planning sessions for all machines per week	no.	75
3.2	Running time calculation for a planning session	min	8
	Set-up time, cycle, number of cavities, mold, pigment		
	Total time per year	h	530
	Hourly pay rate for planners	€/h	30.00
3.3	**Total cost of planning**	**€**	**15,900**
4	**Reduction in scrap levels**		
4.1	Process monitoring, prevention of scrap, option	%	0
4.2	Analysis of scrap, reduction in scrap	%	0.4
4.3	**Total avoidable scrap costs**	**€**	**15,600**
5	**Reduction in management costs for servicing and maintenance**		
5.1	Servicing for machines, molds, peripherals	pcs	98
	Time outlay on monitoring per resource per month	h	0.25
	Hourly rate	€/h	30.00
5.2	**Management reduction per year**	**€**	**8,820**
6	**Summary per year**		
6.1	Total costs of data acquisition and evaluation	**€**	**25,228**
6.2	Total loss of effectivity	**€**	**30,000**
6.3	Total cost of planning	**€**	**15,900**
6.4	Total avoidable scrap costs	**€**	**15,600**
6.5	Reduction in management servicing and maintenance	**€**	**8,820**
6.6	**Total per year**	**€**	**95,548**

Fig. 11.8. Example of an ROI analysis

11.12 Summary

Increasing competitive pressure and high quality requirements are forcing companies in the plastics processing industry to exploit all possible methods of increasing economic efficiency. Here the use of an MES system is one of the most important resources for achieving that. Real-time control of the business processes and of controlling are the central conditions which must be met if the company's market share is to be maintained or expanded. Exact information is needed and must be available in real time. Only an MES system is capable of supplying these data.

That use will be made in the future of the advantages of such systems is also shown by studies conducted by leading market research institutes which forecast two-figure growth rates for the MES market in the next few years.

Abbreviations

APO	Advanced Planner and Optimizer (SAP APO)
BAPI	Business Application Programming Interface
BSA	Business Service Architecture (type of software architecture particularly suitable for business models)
CAQ	Computer-Aided Quality Assurance
DNC	Distributed Numerical Control, Direct Numerical Control
EDI	Electronic Data Interchange (standardized method of exchanging data between businesses)
EDP	Electronic Data Processing
ERP	Enterprise Resource Planning
ESA	Enterprise Solution/System Architecture
ESA	Enterprise Service Architecture (type of software architecture particularly suitable for business models)
FDA	Food and Drug Administration
GAMP	Good Automation Manufacturing Process
IDOC	Intermediate Document (special data format in the SAP environment)
JDBC	Java Database Connectivity (database access method for the Java programming language)
KPI	Key Performance Indicator
MDC	Machine Data Collection
MES	Manufacturing Execution System
MIS	Management Information System
MM	Materials Management
ODBC	Open Database Connectivity (standardized database access method under Microsoft Windows)

OLE	Object Linking and Embedding
OLTP	Online Transaction Processing (procedure for real-time data access; opposite of data warehouse)
OPC	OLE for Process Control (Windows-based interface for data exchange between two systems)
OPC	OLE Process Communication
PDA	Production Data Acquisition
PI-PCS	SAP-interface for Process Control Systems
PM	SAP Maintenance
PP	SAP Production Planning
PP-PDC	Plant Data Collection
PP-POI	Production Optimizing Interface
PPS	Production Planning and control System
QA	Quality Assurance
QM-IDI	Inspection Data Interface
RFC	Remote Function Call (interface for data exchange between two systems)
RFI	Radio Frequency Identification
RFID	Radio Frequency Identification
ROI	Return On Investment
SAP GUI	SAP Graphical User Interface
SAP-HR	SAP personnel management
SAP-QM	SAP Quality Management
SQL	Structured Query Language (query language for relational databases)
TCP/IP	Network protocol for both intranet and internet
VDI	Verein Deutscher Ingenieure
WIP	Work In Process
WLAN	Wireless LAN (network based on radio waves)
XAPP	Cross-Application
XI	Exchange Infrastructure

Checklist

Preliminary note for the user

As a rule it is a difficult task to identify criteria for selecting an MES system – and thus an MES system supplier – and anyone trying to do so can rapidly be overtaxed. The more you understand about the matter, the harder it becomes to keep an overview of it. Where should you start and where should you stop? Once you've looked at different system presentations it is difficult to see where they differ; suddenly all user interfaces look the same.

MES systems are complex IT installations which, depending on the form they take, can affect a large number of functional areas in a manufacturing company. Possibilities for using the systems range from straightforward PDA feedback regarding quality assurance and personnel management to complex detailed scheduling control systems.

When objectives have been defined, a very much better assessment can be made of the scope of services offered by different systems and suppliers. For example, if you want to start with machine data collection but already know that in one or two years the company will be tackling the subject of incentive wages, you can immediately lay down KO criteria. Any potential supplier not able to integrate a staff work time logging function and a performance-based remuneration function can then be ruled out at once.

With our proposal we intend First to point out once again the possible objectives which can be secured by introducing an MES system. This will usually mean that even quantifiable approaches can be found which then form the basis for calculating a return on investment.

This procedure is more efficient and involves less work. In addition, the level to which objectives have been achieved can be better checked for the purpose of subsequent investment controlling.

The following checklist should provide some assistance in designing and selecting MES systems.

It should be taken as a suggested way of systematically evaluating different MES systems and of drawing up selection criteria on this basis.

General criteria

Does the MES system have fully integrated production, personnel and quality management?
Does the MES support paperless (low-paper) production?
Does the MES system include all necessary standard products?
Does the MES offer escalation management and workflow functions?
What references and knowledge of the industry does the supplier have?
How easy is it to adapt the functionalities to the customer's processes?
Does the MES manufacturer have a clear standard product and release strategy?

System concept

Is the complete MES functionality provided in a single system?
Can the individual components be used as modules?
Can the functions be configured?
Does the MES system have an ESA-oriented architecture?
Does the MES system orient itself by common industry standard products?
Does the MES system support the necessary platforms?
Does the MES system support the necessary interfaces?
How easy is it to adapt interfaces to the requirements of the customer?
How easily can the clients be adapted to the requirements of the customer?
What possibilities does the MES system offer for the customer's own developments?
Can these adaptations be made just as easily at a later point in time?
What aids are available for preparing one's own analyses?
Can the existing analyses in various levels of data aggregation be adjusted for all corporate levels?
Is there an interface with leading ERP and PPS systems?
Does the modular architecture of the MES system permit a gradual expansion to include further functions?
Is the system architecture open?

Production

Are there integrated functions which offer a view of all resources involved in production?
Are there overviews which allow evaluation of the current situation?
Are the detailed planning functions based on current PDA data?

Does detailed planning manage primary and secondary resources?
Is there a load planning function for different kinds of secondary resources?
Is it possible to model different possibilities of technological relationships?
Is inter-order networking possible?
Can the types of capacity be varied?
Are different planning strategies supported?
Can detailed planning instances be evaluated by means of flexible and combinable key data?
Can alternative planning variants be simulated?
Can different optimization strategies be implemented?
Does the MES support different production structures (multiple machine work, multiple operator work, and so on)?
Is material tracking possible (in batches and buffer storage, for example)?

Quality

Can quality inspections be incorporated like work operations in an overall order structure?
Is there a dynamic configuration function for monitoring testing and inspection equipment?
Is non-conformance management workflow-supported?
Is gapless traceability of the production process possible?
Does production planning have access to quality data?
Can process (measurement) data also be used as quality characteristics?
Is there support for automatic transfer of measured data via standard interfaces?

Personnel

Is staff work time logging with information and intelligence functions available at the terminal?
Can working hours and payments models be configured simply for personnel time management?
Is there a workflow for paperless processing of applications and approvals?
Is it a straightforward matter to adapt calculation of incentive wages to collective bargaining agreements?
Is there a short-term manpower planning function with direct coupling to production loading?

Is there a short-term manpower planning function which automatically assigns employees on the basis of their qualifications to work centers at which orders have been scheduled?

Data acquisition

Does the MES permit gapless automated data acquisition and processing?
Are standard interfaces provided to machines and automatically controlled machines?
Can all data acquisition functions be integrated in a single data acquisition terminal?
Can data acquisition functions be configured for better ergonomics and thus greater acceptance?
Are standard data acquisition interfaces such as OPC or Euromap 63 supported?
Are the data acquisition functions available on different platforms touch screen terminal, mobile terminal, PC, the web?
Is data acquisition supported by suitable peripheral devices such as ID readers of the most varied types, label printers and so on?

MES in the SAP environment

Does the manufacturer of the MES system have the corresponding SAP know-how?
How high is the number of implementations with SAP manufacturing customers?
Does the MES manufacturer have consultants with SAP application know-how?
Is the manufacturer certificated for application interfaces such as PP-PDC?
Does he have the "Powered by NetWeaver" certificate?
Does the system support the following SAP interfaces?

– Production control	PP-PDC, PP-POI
– Materials management	MM-MOB
– Quality management	QM-IDI
– Production process industry	PI-PCS
– Quantity production	BAPI for plan orders

Can BAPI's be used?

Does the manufacturer have his own in-house SAP development and testing system for customer-specific implementations and support?

Updating

In practice there are continuous further developments in fields such as interfaces, statutory requirements, software tools, requirements emerging from daily activities, and so on

The authors

Bernd Berres, born 1971, C. Eng. (BA) After graduating in technical informatics from the Mosbach Vocational Institute, he became product manager for staff work time logging in MES HYDRA at MPDV Mikrolab GmbH. Also responsible for access control, short-term manpower planning and performance-based wage calculation and their further development.

Otto Brauckmann, born 1938, master's degree in business economics
Studied business economics at the Ludwig-Maximilian-Universität in Munich, majoring in cost accounting.
Since 1984 an independent consultant and distributor for systems for production data acquisition and quality assurance.
In 2002 *Integrated Production Data Management* published by Gabler Verlag. Co-author of *Manufacturing Scorecard* (2004) published by Gabler Verlag.

Andreas Cordt, born 1966, master's degree in statistics from the University of Dortmund majoring in quality management.
Followed by employment in a software company with responsibility for CAQ software sales. Joined MPDV- Mikrolab GmbH in January 2004 specializing in consulting.

Rainer Deisenroth, born 1953, C. Eng. After graduating in technical informatics, worked in hardware and software development in the field of measuring instruments and computer systems, including at Bopp & Reuther GmbH in Mannheim. Following a change to the fields of product management and marketing, joined MPDV Mikrolab GmbH in 1990 where he was responsible for setting up a sales and marketing organization. Currently an authorized company signatory and head of MPDV Sales.

Leonhard Fleischer, born 1964. C. Eng. (BA) Studied mechanical engineering at Mosbach Vocational Institute, majoring in design engineering. Specialized in plastics processing (injection molding).
In 1987 joined MPDV Mikrolab GmbH in Mosbach. Since 1991 head of software development.

Martin Geppert, born 1964, master's degree in information sciences.
Studied information sciences with business management at the University of Karlsruhe (TH).
Master's thesis on knowledge-based planning of production systems.
Since 1991 employed at MPDV Mikrolab GmbH in Mosbach in support organization.
Since 2001 head of the Consulting and Training departments of MPDV Mikrolab GmbH.

Rainer Glatz, master's degree in information sciences from University of Karlsruhe. Since 1981 research assistant at the Institute for Computer Applications in Planning and Design of the University of Karlsruhe.
Since 1987 consultant in the Informatics department at the VDMA in Frankfurt.
Since 1990 head of the Informatics department at the VDMA and since 1999 head of its Software professional association.
Since 2000 head of Industrial Communication professional association.

Torsten Keil, born 1971, electronics engineer. Since 1993 software developer at CS Informatik GmbH. In addition to programming quality management solutions, project manager in system introductions.
Since 2004 head of software development in the quality assurance division at the Stuttgart branch.

Dr. Jürgen Kletti, born 1948, studied electrical engineering majoring in technical data processing at the University of Karlsruhe. After graduation founded MPDV Mikrolab GmbH, of which he is still a partner and CEO today. Since 1990 MPDV has been primarily involved with software products and services for manufacturing industry. MPDV's flagship product is the MES system HYDRA.

Wolfhard Kletti, born 1958, master's degree in information sciences from the Technical Engineering College of Mannheim.
Freelancer at IBM and in projects at the ETH Zürich.
University of Karlsruhe specializing in databases for engineering applications. Joined MPDV Mikrolab GmbH in 1986. Today member of general management with the specialist field of consulting.

Wolfgang Nonnenmann, born 1945, C. Eng. Studied electrical engineering, majoring in closed-loop control systems and electronic data processing at the University of Karlsruhe. Following graduation, worked at Bosch's development center for automotive equipment in Stuttgart.
Founded the company IBN-Systems which specialized in production data acquisition in the fields of plastics and automobiles. Today active as an consultant in the field of production management, PDA and PPS/ERP.

Thorsten Strebel, born 1972, C. Eng. (BA). Studied technical informatics at Mosbach Vocational Institute, majoring in production information systems.
After graduating, management consultant on the implementation of development projects, specializing in object orientation.
Since 1997 senior consultant at MPDV Mikrolab GmbH specializing in production data acquisition, material flow acquisition and detailed planning. Head of the SAP Competence Center.

Index